Outgrowing the Earth

OUTGROWING THE EARTH

The Food Security Challenge in an Age of Falling Water Tables and Rising Temperatures

Lester R. Brown

First published by Earthscan in the UK and USA in 2005

2 Park Square, Milton Park, Abingdon, Oxfordshire OX14 4RN
52 Vanderbilt Avenue, New York, NY 10017

Routledge is an imprint of the Taylor & Francis Group, an informa business

First issued in paperback 2018

ISBN 978-1-84407-185-2 (hbk)
ISBN 978-1-138-38010-3 (pbk)

Typesetting by Elizabeth Docherty

Cover design by Yvonne Booth

A catalogue record for this book is available from the British Library

Acknowledgments

The time between the decision to do this book and its delivery to our publisher, W.W. Norton, was five months. Producing an unscheduled book quickly is possible only if you have an experienced support team like the one I have at the Earth Policy Institute.

This book, like every other book I have written, was dictated. Reah Janise Kauffman, our vice president and my special assistant, transcribed the tapes. She's responsible for producing the manuscript in the early stages and is my first line of feedback, eager to let me know when she thinks ideas and concepts are working.

During our research for *Outgrowing the Earth*, Janet Larsen, my research colleague, used her training in Earth Systems at Stanford to help me think through some of the complex issues involved in analyzing food security. She was also my most comprehensive critic, always steering me toward a stronger, clearer, more concise book.

Underpinning the production of this book is a database drawn from many fields by our graphic data organizer Viviana Jiménez. Viviana helped with research from the beginning and followed up with fact checking. Lila Buckley joined our staff as an intern just in time to provide welcome support and a fresh perspective as we started the review process.

Once a book is finished, our thoughts turn to marketing. Millicent Johnson, who manages the sales database and thousands of book orders, takes pride in her one-day turnaround policy for incoming orders.

In addition to managing the Institute so I can concentrate on research, Reah Janise also manages our worldwide publishing network. She, more than anyone, is responsible for the publication of my various books in more than 40 languages. A list of foreign translations of our books is available at www.earth-policy.org/Books/intl.htm.

The researching, writing, editing, and publishing of a book is much easier when you have an experienced team at every step of the way. Our editor, Linda Starke, brings 30 years of experience in editing environmental books and reports. She has brought her sure hand to the editing of not only this book, but more than 40 other books we have done together, including some 18 *State of the World* reports and 10 annual editions of *Vital Signs* when I was at the Worldwatch Institute. Reah Janise has worked with me on over 35 books. For Janet, this is our fourth book together, including one she coauthored. And for Viviana, our second.

Production of the book in record time is only possible thanks to the efforts of Elizabeth Doherty, who gave up her evenings and weekends to prepare the page proofs. The index has again been ably prepared under deadline pressure by Ritch Pope.

My relationship with W.W. Norton & Company, a marriage made in heaven, now includes some 50 titles. My thanks to the team at Norton: Amy Cherry, our editor; Lucinda Bartley, her assistant; Andrew Marasia and Amanda Morrison, who put the book on a fast-track production schedule; Ingsu Liu, Art Director for the book jacket; and Bill Rusin, Marketing Director. It is a delight to work with such a talented team.

As with any book, reviewers help shape the final product. Toby Clark, a consummate environmental professional, brought his keen insights and decades of experience in environmental policy at the Environmental Protection Agency and the Council on Environmental Quality to bear on the manuscript. Maureen Kuwano Hinkle reviewed the evolving manuscript twice, drawing on her 26 years of experience working on agricultural issues with Environmental Defense and the Audubon Society.

Kenneth Cassman at the University of Nebraska used his wealth of experience in assessing crop yield potential to help strengthen and fine-tune several agronomic points. Bill Mansfield, of our Board, also read the manuscript for us, providing general feedback. From the staff, Janet, Reah Janise, Viviana, and Lila each read it twice, providing helpful feedback as the book evolved.

We are indebted to an extraordinarily supportive Board of Directors, chaired by Judy Gradwohl. In addition to Judy and Bill, our other Board members are Scott McVay, Raisa Scriabine, and Hamid Taravati.

Earth Policy is supported by a network of dedicated publishers for our books and *Eco-Economy Updates* in 22 languages—Arabic, Catalan, Chinese, Czech, Danish, English, French, Indonesian, Italian, Japanese, Korean, Marathi (India), Persian, Polish, Portuguese (in Brazil), Romanian, Russian, Spanish, Swedish, Thai, Turkish, and Ukrainian. There are three editions in English (U.S.A./Canada, U.K./Commonwealth, and India/South Asia), and two in Spanish (Spain and Latin America). We have benefited from strong support from a number of individuals responsible for various publishing arrangements—going back, in some cases, 20 or more years—

including Gianfranco Bologna in Italy, Soki Oda in Japan, Lin Zixin in China, Hamid Taravati and Farzaneh Bahar in Iran, Jonathan Sinclair Wilson in the United Kingdom, President Ion Iliescu and Roman Chirila in Romania, and, more recently, Eduardo Athayde in Brazil and Nandini Rao in India.

And finally, but most important, this book would not have been possible were it not for the generous support of funders. Among these are the U.N. Population Fund and several foundations, including the Appleton, Fred Gellert Family, Richard and Rhoda Goldman, Farview, McBride Family, Shenandoah, Summit, Turner, and Wallace Genetic foundations. Their support is invaluable to our work of researching and disseminating the vision of an eco-economy. We would also like to thank Junko Edahiro, Susan Brown, Judy Hyde, Leonora Barnheisel, and Peter Seidel for their generous individual contributions to the Institute.

Lester R. Brown

Contents

Preface

On hearing his political opponent described as a modest chap, Winston Churchill reputedly responded that "he has much to be modest about." Having just completed a book dealing with the increasingly complex issue of world food security, I too feel that I have a lot to be modest about.

Assessing the world food prospect was once rather straightforward, largely a matter of extrapolating, with minor adjustments, historically recent agricultural supply and demand trends. Now suddenly that is all changing. It is no longer just a matter of trends slowing or accelerating; in some cases they are reversing direction.

Grain harvests that were once rising everywhere are now falling in some countries. Fish catches that were once rising are now falling. Irrigated area, once expanding almost everywhere, is now shrinking in some key food-producing regions.

Beyond this, some of the measures that are used to expand food production today, such as overpumping aquifers, almost guarantee a decline in food production tomorrow when the aquifers are depleted and the wells go dry. The same can be said for overplowing and overgrazing. We have entered an era of discontinuity on the food front, an era where making reliable projections is ever more difficult.

New research shows that a 1 degree Celsius rise in temperature leads to a decline in wheat, rice, and corn yields of 10 percent. In a century where temperatures could rise by several degrees Celsius, harvests could be devastated.

Although climate change is widely discussed, we are slow to grasp its full meaning. Everyone knows the earth's temperature is rising, but commodity analysts often condition their projections on weather returning to "normal," failing to realize that with climate now in flux, there is no normal to return to.

Falling water tables are also undermining food security. Water tables are now falling in countries that contain more than half the world's people. While there is a broad realization that we are facing a future of water shortages, not everyone has connected the dots to see that a future of water shortages will be a future of food shortages.

Perhaps the biggest agricultural reversal in recent times has been the precipitous decline in China's grain production since 1998. Ten years ago, in *Who Will Feed China?*, I projected that China's grain production would soon peak and begin to decline. But I did not anticipate that it would drop by 50 million tons between 1998 and 2004. Since 1998 China has covered this decline by drawing down its once massive stocks of grain. Now stocks are largely depleted and China is turning to the world market. Its purchase of 8 million tons of wheat to import in 2004 could signal the beginning of a shift from a world food economy dominated by surpluses to one dominated by scarcity.

Overnight, China has become the world's largest wheat importer. Yet it will almost certainly import even more wheat in the future, not to mention vast quantities of rice and corn. It is this potential need to import 30, 40, or 50 million tons of grain a year within the next year or two and the associated emergence of a politics of food scarcity that is likely to put food security on the front page of newspapers.

At the other end of the spectrum is Brazil, the only country with the potential to expand world cropland area measurably. But what will the environmental consequences be of continuing to clear and plow Brazil's vast interior? Will the soils sustain cultivation over the longer term? Will the deforestation in the Amazon disrupt the recycling of rainfall from the Atlantic Ocean to the country's interior? And how many plant and animal species will Brazil sacrifice to expand its exports of soybeans?

Food security, which was once the near-exclusive province of ministries of agriculture, now directly involves several departments of government. In the past, ministries of transportation did not need to think about food security when formulating transport policies. But in densely populated developing countries today, the idea of having a car in every garage one day means paving over a large share of their cropland. Many countries simply do not have enough cropland to pave for cars and to grow food for their people.

Or consider energy. Energy ministers do not attend international conferences on food security. But they should. The decisions they make in deciding which energy sources to develop will directly affect atmospheric carbon dioxide levels and future changes in temperature. In fact, the decisions made in ministries of energy may have a greater effect on long-term food security than those made in ministries of agriculture.

Future food security now depends on the combined efforts of the ministries of agriculture, energy, transportation, health and family planning, and water resources. It also depends on strong leadership—leadership that is far better informed on the complex set of interacting forces affecting food security than most political leaders are today.

Lester R. Brown

Earth Policy Institute
1350 Connecticut Ave., NW, Suite 403
Washington, DC 20036

Phone: (202) 496-9290
Fax: (202) 496-9325
E-mail: epi@earth-policy.org
Web: www.earth-policy.org

October 2004

Data for figures and additional information can be found at www.earth-policy.org/Books/Out/index.htm.

Outgrowing the Earth

1

Pushing Beyond the Earth's Limits

When historians look back on our times, the last half of the twentieth century will undoubtedly be labeled "the era of growth." Take population. In 1950, there were 2.5 billion people in the world. By 2000, there were 6 billion. There has been more growth in world population since 1950 than during the preceding 4 million years.[1]

Recent growth in the world economy is even more remarkable. During the last half of the twentieth century, the world economy expanded sevenfold. Most striking of all, the growth in the world economy during the single year of 2000 exceeded that of the entire nineteenth century. Economic growth, now the goal of governments everywhere, has become the status quo. Stability is considered a departure from the norm.[2]

As the economy grows, its demands are outgrowing the earth, exceeding many of the planet's natural capacities. While the world economy multiplied sevenfold in just 50 years, the earth's natural life-support systems remained essentially the same. Water use tripled, but the capacity of the hydrological system to produce fresh water through evaporation changed little. The demand for seafood increased fivefold, but the sustainable yield of oceanic fisheries was unchanged. Fossil fuel burning raised carbon dioxide (CO_2) emissions fourfold, but the capacity of nature to absorb CO_2 changed little, leading to a buildup of CO_2 in the atmosphere and a rise in the earth's temperature. As human demands surpass the earth's natural capacities, expanding food production becomes more difficult.[3]

Losing Agricultural Momentum

Environmentalists have been saying for years that if the environmental trends of recent decades continued the world would one day be in trouble. What was not clear was what form the trouble would take and when it would occur. It now seems likely to take the form of tightening food supplies, and within the next few years. Indeed, China's forays into the world market in early 2004 to buy 8 million tons of wheat could mark the beginning of the global shift from an era of grain surpluses to one of grain scarcity.[4]

World grain production is a basic indicator of dietary adequacy at the individual level and of overall food security at the global level. After nearly tripling from 1950 to 1996, the grain harvest stayed flat for seven years in a row, through 2003, showing no increase at all. And in each of the last four of those years, production fell short of consumption. The shortfalls of nearly 100 million tons in 2002 and again in 2003 were the largest on record.[5]

With consumption exceeding production for four years, world grain stocks dropped to the lowest level in 30 years. (See Figure 1–1.) The last time stocks were this low, in 1972–74, wheat and rice prices doubled. Importing countries competed vigorously for inadequate supplies. A politics of scarcity emerged—with some countries, such as the United States, restricting exports.[6]

In 2004 a combination of stronger grain prices at planting time and the best weather in a decade yielded a

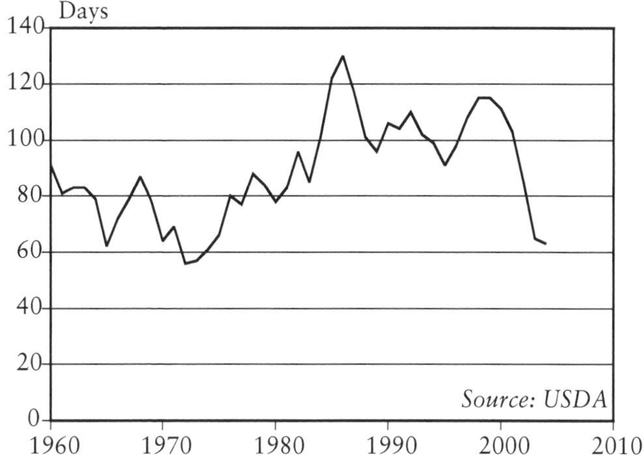

Figure 1–1. *World Grain Stocks as Days of Consumption, 1960–2004*

substantially larger harvest for the first time in eight years. Yet even with a harvest that was up 124 million tons from that in 2003, the world still consumed all the grain it produced, leaving none to rebuild stocks. If stocks cannot be rebuilt in a year of exceptional weather, when can they?[7]

From 1950 to 1984 world grain production expanded faster than population, raising the grain produced per person from 250 kilograms to the historical peak of 339 kilograms, an increase of 34 percent. This positive development initially reflected recovery from the disruption of World War II, and then later solid technological advances. The rising tide of food production lifted all ships, largely eradicating hunger in some countries and substantially reducing it in many others.[8]

Since 1984, however, grain harvest growth has fallen behind that of population, dropping the amount of grain produced per person to 308 kilograms in 2004, down 9 percent from its historic high point. Fortunately, part of the global decline was offset by the increasing efficiency with which feedgrains are converted into animal protein, thanks to the growing use of soybean meal as a protein supplement. Accordingly, the deterioration in nutrition has not been as great as the bare numbers would suggest.[9]

The one region where the decline in grain produced per person is unusually steep and where it is taking a heavy human toll is Africa. In addition to the nutrient depletion of soils and the steady shrinkage in grainland per person from population growth in recent decades, Africa must now contend with the loss of adults to AIDS, which is depleting the rural work force and undermining agriculture. From 1960 through 1981, grain production per person in sub-Saharan Africa ranged between 140 and 160 kilograms per person. (See Figure 1–2.) Then from 1980 through 2001 it fluctuated largely between 120

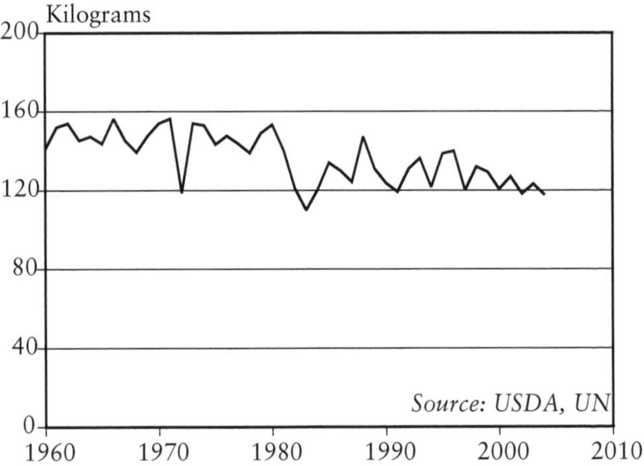

Figure 1–2. *Grain Production Per Person in Sub-Saharan Africa, 1960–2004*

and 140 kilograms. And in two of the last three years, it has been below 120 kilograms—dropping to a level that leaves millions of Africans on the edge of starvation.[10]

Several long-standing environmental trends are contributing to the global loss of agricultural momentum. Among these are the cumulative effects of soil erosion on land productivity, the loss of cropland to desertification, and the accelerating conversion of cropland to nonfarm uses. All are taking a toll, although their relative roles vary among countries.

Now two newer environmental trends—falling water tables and rising temperatures—are slowing the growth in world food production, as described later in this chapter. (See also Chapters 6 and 7.) In addition, farmers are faced with a shrinking backlog of unused technology. The high-yielding varieties of wheat, rice, and corn that were developed a generation or so ago are now widely used in industrial and developing countries alike. They doubled and tripled yields, but there have not been any dramatic advances in the genetic yield potential of grains since then.[11]

The use of fertilizer, which removed nutrient constraints and helped the new high-yielding varieties realize their full genetic potential during the last half-century, has now plateaued or even declined slightly in key food-producing countries. Among these are the United States, countries in Western Europe, Japan, and now possibly China as well. Meanwhile, the rapid growth in irrigation that characterized much of the last half-century has also slowed. Indeed, in some countries the irrigated area is shrinking.[12]

The bottom line is that it is now more difficult for farmers to keep up with the growing demand for grain. The rise in world grainland productivity, which averaged over 2 percent a year from 1950 to 1990, fell to scarcely

1 percent a year from 1990 to 2000. This will likely drop further in the years immediately ahead.[13]

If the rise in land productivity continues to slow and if population continues to grow by 70 million or more per year, governments may begin to define national security in terms of food shortages, rising food prices, and the emerging politics of scarcity. Food insecurity may soon eclipse terrorism as the overriding concern of national governments.[14]

Growth: The Environmental Fallout

The world economy, as now structured, is making excessive demands on the earth. Evidence of this can be seen in collapsing fisheries, shrinking forests, expanding deserts, rising CO_2 levels, eroding soils, rising temperatures, falling water tables, melting glaciers, deteriorating grasslands, rising seas, rivers that are running dry, and disappearing species.

Nearly all these environmentally destructive trends adversely affect the world food prospect. For example, even a modest rise of 1 degree Fahrenheit in temperature in mountainous regions can substantially increase rainfall and decrease snowfall. The result is more flooding during the rainy season and less snowmelt to feed rivers during the dry season, when farmers need irrigation water.[15]

Or consider the collapse of fisheries and the associated leveling off of the oceanic fish catch. During the last half-century the fivefold growth in the world fish catch that satisfied much of the growing demand for animal protein pushed oceanic fisheries to their limits and beyond. Now, in this new century, we cannot expect any growth at all in the catch. All future growth in animal protein supplies can only come from the land, putting even more pressure on the earth's land and water resources.[16]

Farmers have long had to cope with the cumulative effects of soil erosion on land productivity, the loss of cropland to nonfarm uses, and the encroachment of deserts on cropland. Now they are also being battered by higher temperatures and crop-scorching heat waves. Likewise, farmers who once had assured supplies of irrigation water are now forced to abandon irrigation as aquifers are depleted and wells go dry. Collectively this array of environmental trends is making it even more difficult for farmers to feed adequately the 70 million people added to our ranks each year.[17]

Until recently, the economic effects of environmental trends, such as overfishing, overpumping, and overplowing, were largely local. Among the many examples are the collapse of the cod fishery off Newfoundland from overfishing that cost Canada 40,000 jobs, the halving of Saudi Arabia's wheat harvest as a result of aquifer depletion, and the shrinking grain harvest of Kazakhstan as wind erosion claimed half of its cropland.[18]

Now, if world food supplies tighten, we may see the first global economic effect of environmentally destructive trends. Rising food prices could be the first economic indicator to signal serious trouble in the deteriorating relationship between the global economy and the earth's ecosystem. The short-lived 20-percent rise in world grain prices in early 2004 may turn out to be a warning tremor before the quake.[19]

Two New Challenges

As world demand for food has tripled, so too has the use of water for irrigation. As a result, the world is incurring a vast water deficit. But because this deficit takes the form of aquifer overpumping and falling water tables, it is nearly invisible. Falling water levels are often not discovered until wells go dry.[20]

The world water deficit is historically recent. Only within the last half-century, with the advent of powerful diesel and electrically driven pumps, has the world had the pumping capacity to deplete aquifers. The worldwide spread of these pumps since the late 1960s and the drilling of millions of wells, mostly for irrigation, have in many cases pushed water withdrawal beyond the aquifer's recharge from rainfall. As a result, water tables are now falling in countries that are home to more than half of the world's people, including China, India, and the United States—the three largest grain producers.[21]

Groundwater levels are falling throughout the northern half of China. Under the North China Plain, they are dropping one to three meters (3–10 feet) a year. In India, they are falling in most states, including the Punjab, the country's breadbasket. And in the United States, water levels are falling throughout the southern Great Plains and the Southwest. Overpumping creates a false sense of food security: it enables us to satisfy growing food needs today, but it almost guarantees a decline in food production tomorrow when the aquifer is depleted.[22]

With 1,000 tons of water required to produce 1 ton of grain, food security is closely tied to water security. Seventy percent of world water use is for irrigation, 20 percent is used by industry, and 10 percent is for residential purposes. As urban water use rises even as aquifers are being depleted, farmers are faced with a shrinking share of a shrinking water supply.[23]

At the same time that water tables are falling, temperatures are rising. As concern about climate change has intensified, scientists have begun to focus on the precise relationship between temperature and crop yields. Crop ecologists at the International Rice Research Institute in the Philippines and at the U.S. Department of Agriculture (USDA) have jointly concluded that with each 1-degree

Celsius rise in temperature during the growing season, the yields of wheat, rice, and corn drop by 10 percent.[24]

Over the last three decades, the earth's average temperature has climbed by nearly 0.7 degrees Celsius, with the four warmest years on record coming during the last six years. In 2002, record-high temperatures and drought shrank grain harvests in both India and the United States. In 2003, it was Europe that bore the brunt of the intense heat. The record-breaking August heat wave that claimed 35,000 lives in eight nations withered grain harvests in virtually every country from France in the west through the Ukraine in the east.[25]

The Intergovernmental Panel on Climate Change projects that during this century, with a business-as-usual scenario, the earth's average temperature will rise by 1.4–5.8 degrees Celsius (2–10 degrees Fahrenheit). These projections are for the earth's average temperature, but the rise is expected to be much greater over land than over the oceans, in the higher latitudes than in the equatorial regions, and in the interior of continents than in the coastal regions. This suggests that increases far in excess of the projected average are likely for regions such as the North American breadbasket—the region defined by the Great Plains of the United States and Canada and the U.S. Corn Belt. Today's farmers face the prospect of temperatures higher than any generation of farmers since agriculture began.[26]

The Japan Syndrome

When studying the USDA world grain database more than a decade ago, I noted that if countries are already densely populated when they begin to industrialize rapidly, three things happen in quick succession to make them heavily dependent on grain imports: grain consumption climbs as incomes rise, grainland area shrinks, and grain production falls. The rapid industrialization that drives up demand simultaneously shrinks the cropland area. The inevitable result is that grain imports soar. Within a few decades, countries can go from being essentially self-sufficient to importing 70 percent or more of their grain. I call this the "Japan syndrome" because I first recognized this sequence of events in Japan, a country that today imports 70 percent of its grain.[27]

In a fast-industrializing country, grain consumption rises rapidly. Initially, rising incomes permit more direct consumption of grain, but before long the growth shifts to the greater indirect consumption of grain in the form of grain-intensive livestock products, such as pork, poultry, and eggs.

Once rapid industrialization is under way, it is usually only a matter of years before the grainland area begins to shrink. Among the trends leading to this are the abandonment of marginal cropland, the loss of rural labor needed for multiple cropping, and a shift of grainland to the production of fruits, vegetables, and other high-value crops.

First, as a country industrializes and modernizes, cropland is used for industrial and residential developments. As automobile ownership spreads, the construction of roads, highways, and parking lots also takes valuable land away from agriculture. In situations where farmers find themselves with fragments of land that are too small to be economically cultivated, they often simply abandon their plots, seeking employment elsewhere.

Second, as rapid industrialization pulls labor out of the countryside, it often leads to less double cropping, a practice that depends on quickly harvesting one grain crop once it is ripe and immediately preparing the seedbed for the next crop. With the loss of workers as young people migrate to cities, the capacity to do this diminishes.

Third, as incomes rise, diets diversify, generating

demand for more fruits and vegetables. This in turn leads farmers to shift land from grain to these more profitable, high-value crops.

Japan was essentially self-sufficient in grain when its grain harvested area peaked in 1955. Since then the grain-land area has shrunk by more than half. The multiple-cropping index has dropped from nearly 1.4 crops per hectare per year in 1960 to scarcely 1 today. Some six years after Japan's grain area began to shrink, the shrinkage overrode the rise in land productivity and overall production began to decline. With grain consumption climbing and production falling, grain imports soared. (See Figure 1–3.) By 1983 imports accounted for 70 percent of Japan's grain consumption, a level they remain at today.[28]

A similar analysis for South Korea and Taiwan shows a pattern that is almost identical with that of Japan. In both cases, the decline in grain area was followed roughly a decade later by a decline in production. Perhaps this

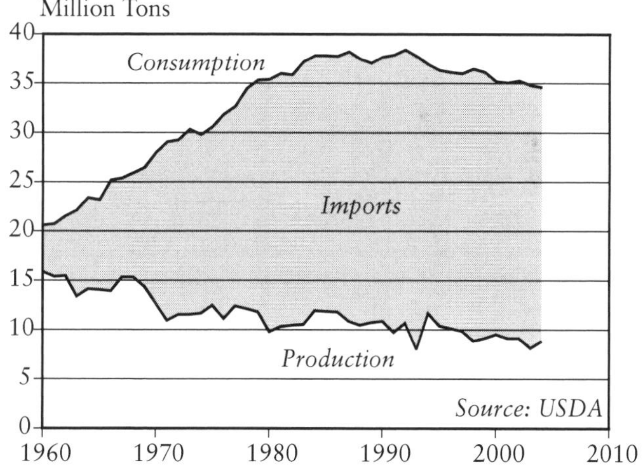

Figure 1–3. *Grain Production, Consumption, and Imports in Japan, 1960–2004*

should not be surprising, since the forces at work in the two countries are exactly the same as in Japan. And, like Japan, both South Korea and Taiwan now import some 70 percent of their total grain supply.[29]

Based on the sequence of events in these three countries that affected grain production, consumption, and imports—the Japan syndrome—it was easy to anticipate the precipitous decline in China's grain production that began in 1998 (as described in the next section). The obvious question now is which other countries will enter a period of declining grain production because of the same combination of forces? Among those that come to mind are India, Indonesia, Bangladesh, Pakistan, Egypt, and Mexico.[30]

Of particular concern is India, home to nearly 1.1 billion people. In recent years, its economic growth has accelerated, averaging 6–7 percent a year. This growth, only slightly slower than that of China, is also beginning to consume cropland. So, too, are the needs of the 18 million people added each year to India's population. In addition to the grainland shrinkage associated with the Japan syndrome, the extensive overpumping of aquifers in India—which will one day deprive farmers of irrigation water—will also reduce grain production.[31]

Exactly when rapid industrialization in a country that is densely populated will translate into a decline in grain production is difficult to anticipate. Once production turns downward, countries often try to reverse the trend. But the difficulty of achieving this can be seen in Japan, where a rice support price that is four times the world market price has failed to expand production.[32]

The China Factor

China—the largest country in the world—is now beginning to experience the Japan syndrome. Perhaps the most alarming recent world agricultural event is the precipitous fall in China's grain production since 1998. After an impressive climb from 90 million tons in 1950 to a peak of 392 million tons in 1998, China's grain harvest fell in four of the next five years, dropping to 322 million tons in 2003. For perspective, this decline of 70 million tons exceeds the entire grain harvest of Canada.[33]

Behind this harvest shrinkage of 18 percent from 1998 to 2003 is a decline in grain harvested area of 16 percent. The conversion of cropland to nonfarm uses, the shift of grainland to higher-value fruits and vegetables, and, in some of the more prosperous regions, a loss of the rural labor needed for multiple cropping are all shrinking China's grainland—just as they did Japan's.[34]

In addition, China is also losing grainland to the expansion of deserts and the loss of irrigation water, due to both aquifer depletion and diversion of water to cities. (See Chapter 8 for further discussion of these pressures.) Unfortunately for China, none of the forces that are shrinking the grainland area are easily countered.

Between 1998 and 2003, five consecutive harvest shortfalls dropped China's once massive stocks of grain to their lowest level in 30 years. With stocks now largely depleted, China's leaders—all of them survivors of the great famine of 1959–61, when 30 million people starved to death—are worried. For them, food security is not a trivial issue.[35]

Not surprisingly, China desperately wants to reverse the recent fall in grain production. In March 2004, Beijing announced an emergency supplemental appropriation, expanding the 2004 agricultural budget by one fifth ($3.6 billion) in an effort to encourage farmers to grow more grain. The support price for the early rice crop in 2004 was raised by 21 percent. While these two emergency measures did reverse the grain harvest decline temporarily, whether they can reverse the trend over the longer term is doubtful.[36]

When China turns to the outside world for commodities, it can overwhelm world markets. For example, 10 years ago China was self-sufficient in soybeans. In 2004, it imported 22 million tons—quickly eclipsing Japan, the previous leading importer with 5 million tons.[37]

When wheat prices within China started climbing in the fall of 2003, the government dispatched wheat-buying delegations to Australia, Canada, and the United States. They purchased 8 million tons, and overnight China became the world's largest wheat importer.[38]

China is a fascinating case study because of its sheer size and extraordinary pace of industrial development. It has been the world's fastest-growing economy since 1980. The economic effects of this massive expansion can be seen in the rest of the world, but China is also putting enormous pressure on its own natural resource base. In the deteriorating relationship between the global economy and the earth's ecosystem, China is unfortunately on the cutting edge.[39]

With water, the northern half of China is literally drying out. Water tables are falling, rivers are going dry, and lakes are disappearing. In a 748-page assessment of China's water situation, the World Bank sounds the alarm. It foresees "catastrophic consequences for future generations" if water use and supply cannot quickly be brought back into balance. More immediately, if China cannot quickly restore a balance between the consumption of water and the sustainable yield of its aquifers and rivers, its grain imports will likely soar in the years ahead.[40]

For people not living in China, it is difficult to visualize how fast deserts are expanding. It can be likened to a war, yet it is not invading armies that are claiming the territory, but expanding deserts. Old deserts are advancing and new ones are forming, like guerrilla forces striking unexpectedly, forcing Beijing to fight on several fronts. Throughout northern and western China, some 24,000 villages have either been abandoned or partly depopulated as drifting sand has made farming untenable.[41]

On the food front, the issue within China is not hunger and starvation, as the nation now has a substantial cushion between consumption levels and minimal nutrition needs. Rather, the concern is rising food prices and the effect that this could have on political stability. China's leaders are striving for a delicate balance between food prices that will encourage production in the countryside but maintain stability in the cities.[42]

As noted earlier, smaller countries like Japan, South Korea, and Taiwan can import 70 percent or more of their grain, but if China turns to the outside world to meet even 20 percent of its grain needs, which would be close to 80 million tons, it will provide a huge challenge for grain exporters. The resulting rise in world grain prices could destabilize governments in low-income, grain-importing countries. The entire world thus has a stake in China's efforts to stabilize its agricultural resource base.[43]

The Challenge Ahead

It is difficult to overestimate the challenges the world faces over the next half-century. Not only are there a projected 3 billion more people to feed, but there are also an estimated 5 billion people who want to diversify their diets by moving up the food chain, eating more grain-intensive livestock products. On the supply side, the world's farmers must contend with traditional challenges, such as soil erosion and the loss of cropland to nonfarm uses, but now also with newer trends such as falling water tables, the diversion of irrigation water to cities, and rising temperatures.[44]

At the World Food Summit in 1996 in Rome, 185 governments plus the European Community agreed that the number of hungry people needed to be reduced by half by 2015. Between 1990–92 and 1995–97, the number did decline by some 37 million from 817 million to 780 million, or over 7 million a year—but this was much less than the 20 million per year needed to reach the 2015 target. And then things got even worse. From 1995–97 to 1999–2001, the number of hungry people in the world began to increase, rising by 18 million to 798 million. This increase in hunger is not too surprising, given the lack of growth in the world grain harvest from 1996 to 2003.[45]

Against this backdrop of a slowly deteriorating food situation, there is the prospect that the Japan syndrome will soon take effect in other countries, shrinking their grain harvests. Is India's grain production likely to peak and start declining in the next few years, much as China's did after 1998? Or will India be able to hold off the loss of cropland to nonfarm uses and the depletion of aquifers long enough to eradicate most of its hunger? There are signs that the shrinkage in its grain area, which is a precursor to the shrinkage of overall production, may have begun.

Because aquifer depletion is recent, it is taking agricultural analysts into uncharted territory. It is clear, for example, that water tables are falling simultaneously in many countries and at an accelerating rate. Less clear is exactly when aquifers will be depleted and precisely how much this will reduce food production.

If the climate models projecting the effect of rising atmospheric CO_2 levels on the earth's temperature are anywhere near the mark, we are facing a future of higher temperatures. We do not know exactly how fast temperatures will rise, but in a world of rising temperatures, there is added reason to be concerned about world food security.[46]

On another front, in Africa the spread of HIV/AIDS is threatening the food security of the entire continent as the loss of able-bodied field workers shrinks harvests. In sub-Saharan Africa, disease begets hunger and hunger begets disease. In some villages, high HIV infection rates have claimed an entire generation of young adults, leaving only the elderly and children. Without a major intervention from the outside world, the continuing spread of the virus and hunger that is cutting life expectancy in half in some countries could take Africa back to the Dark Ages.[47]

In a world where the food economy has been shaped by an abundance of cheap oil, tightening world oil supplies will further complicate efforts to eradicate hunger. Modern mechanized agriculture requires large amounts of fuel for tractors, irrigation pumps, and grain drying. Rising oil prices may soon translate into rising food prices.

As we look at the prospect of swelling grain imports for Asia, where half the world's people live, and for Africa, the second most populous continent, we have to ask where the grain will come from. The countries that dominated world grain exports for the last half-century—the United States, Canada, Australia, and Argentina—may not be able to export much beyond current levels.[48]

U.S. grain production, though it has reached 350 million tons several times over the last two decades, has never risen much beyond this. U.S. grain exports, which two decades ago were running around 100 million tons a year, have averaged only 80 million tons in recent years as rising domestic grain use has more than absorbed any production gains. The potential for expansion in both

Canada and Australia is constrained by relatively low rainfall in their grain-growing regions. Argentina's grain production has actually declined over the last several years as land has shifted to soybeans.[49]

By contrast, countries such as Russia and the Ukraine—where population has stabilized or is declining and where there is some unrealized agricultural production potential—should be able to expand their grain exports at least modestly. However, the low yields that are characteristic of northerly countries that depend heavily on spring wheat, as Russia does, will likely prevent Russia from becoming a major grain exporter. The Ukraine has a somewhat more promising potential if it can provide farmers with the economic incentives they need to expand production. So, too, do Poland and Romania.[50]

Yet the likely increases in exports from these countries are small compared with the prospective import needs of China and, potentially, India. It is worth noting that the drop in China's grain harvest of 70 million tons over five years is equal to the grain exports of Canada, Australia, and Argentina combined.[51]

Argentina can expand its already large volume of soybean exports, but its growth potential for grain exports is limited by the availability of arable land. The only country that has the potential to substantially expand the world grainland area is Brazil with its vast *cerrado*, a savannah-like region that lies on the southern edge of the Amazon Basin. (See Chapter 9.) Because its soils require the heavy use of fertilizer and because transporting grain from Brazil's remote interior to distant world markets is costly, it would likely take substantially higher world grain prices for Brazil to emerge as a major exporter. Beyond this, would a vast expansion of cropland in Brazil's interior be sustainable? Or is its vulnerability to soil erosion likely to prevent it from making a long-term contribution? And what will be the price paid in the irretrievable loss of ecosystems and plant and animal species?[52]

Ensuring future food security is a formidable challenge. Can we check the HIV epidemic before it so depletes Africa's adult population that starvation stalks the land? Can we arrest the steady shrinkage in grainland area per person, eliminate the overgrazing that is converting grasslands to desert, and reduce soil erosion losses below the natural rate of new soil formation? Can we simultaneously halt the advancing deserts that are engulfing cropland, check the rising temperature that threatens to shrink harvests, arrest the fall in water tables, and protect cropland from careless conversion to nonfarm uses?

Data for figures and additional information can be found at www.earth-policy.org/Books/Out/index.htm.

2

Stopping at Seven Billion

In early 2003, U.N. demographers announced that the HIV/AIDS epidemic has reduced life expectancy for the 700 million people of sub-Saharan Africa from 62 to 46 years. For the first time in the modern era, the rise in life expectancy has been reversed for a large segment of humanity, marking a major setback in the march of progress. Is this an isolated development? Or does this reversal mark the beginning of a new era where the failure of societies to manage other life-threatening trends, such as falling water tables and rising temperatures, will also disrupt progress and reduce life expectancy?[1]

Over the last three decades, some 35 European countries and Japan have reduced fertility and achieved population stability. Indeed, in many of these countries population is projected to decline somewhat over the next half-century. In all these cases population growth ceased because rising living standards and expanding opportunities for women were reducing births. But now populations are projected to decline in some countries for the wrong reason. In countries with the highest HIV infection rates—Botswana, South Africa, and Swaziland—rising death rates are projected to shrink populations in the decades ahead.[2]

After peaking at an all-time high of 2 percent in 1970, world population growth slowed to 1.2 percent in 2004. This is the good news. The bad news is that part of the slowdown has come from more deaths, mostly from AIDS. Perhaps more important, even slower-growing populations are still outstripping the carrying capacity of the earth's natural systems—its fisheries, forests, rangelands, aquifers, and croplands. Once the demands of a growing population surpass the sustainable yield threshold of an ecosystem, any growth in human numbers is a matter of concern. For example, whether the population-driven demand on a fishery exceeds the sustainable yield by 1 percent or 10 percent a year makes little difference over the long term. The end result is the same: depletion of stocks and collapse of the fishery.[3]

For some areas, population growth now threatens food security. In developing countries, land holdings are parceled out among heirs with each successive generation until they are so small that they can no longer feed a family. The pressure of a larger population can mean a shrinking water supply, leading to hydrological poverty—a situation where there is no longer enough water to drink, to produce food, and for bathing. The continuing growth of population in resource-scarce, low-income countries is undermining future food security in many of them.[4]

A New Demographic Era

Nearly 3 billion people are expected to be added to our world during the first half of this century—slightly fewer than the 3.5 billion added during the last half of the twentieth century. There are some important differences in these numbers, however. Whereas the growth in 1950–2000 occurred in both industrial and developing countries, the growth in the next 50 years will be almost entirely in the developing ones. Big additions are projected for the Indian subcontinent and sub-Saharan Africa, which together will account for nearly 2 billion of the 3 billion total increase.[5]

As noted, populations are projected to shrink in some developing countries, but for the wrong reasons. Whereas the populations of Russia, Japan, and Germany are projected to decline by 2050 by 30, 13, and 3 percent, respectively, due to falling fertility, those of Botswana, South Africa, and Swaziland are expected to decline by 43, 11, and 2 percent because of rising mortality. Are these three African countries an aberration or are they merely among the first of many countries where HIV/AIDS, spreading hunger, the loss of water supplies, and possibly civil conflict lead to rising death rates and population decline?[6]

Another major shift will come as record variations of national population growth and decline redraw the world demographic map. A comparison of the 20 most populous countries in 2000 and those projected for 2050 illustrates these changes. (See Table 2–1.) To begin with, the two largest countries—China and India—will trade places as India's population, projected to grow by over 500 million by 2050, overtakes that of China sometime around 2040.[7]

In the four most populous industrial countries after the United States—Russia, Japan, Germany, and the United Kingdom—populations are projected to be smaller in 2050 than they are today. Indeed, only Japan and Russia will remain among the top 20 by mid-century. Germany and the United Kingdom will drop off the list, as will Thailand, a developing country that is approaching population stability.[8]

The three countries on the list with the greatest growth, with each expected to more than double by 2050, are Pakistan, Nigeria, and Ethiopia. The three newcom-

Table 2–1. *The World's 20 Most Populous Countries, 2000 and 2050*

2000		2050	
Country	Population (million)	Country	Population (million)
China	1,275	India	1,531
India	1,017	China	1,395
United States	285	United States	409
Indonesia	212	Pakistan	349
Brazil	172	Indonesia	294
Russia	146	Nigeria	258
Pakistan	143	Bangladesh	255
Bangladesh	138	Brazil	233
Japan	127	Ethiopia	171
Nigeria	115	Dem. Rep. of the Congo	152
Mexico	99	Mexico	140
Germany	82	Egypt	127
Viet Nam	78	Philippines	127
Philippines	76	Viet Nam	118
Iran	66	Japan	110
Egypt	68	Iran	105
Turkey	68	Uganda	103
Ethiopia	66	Russia	102
Thailand	61	Turkey	98
United Kingdom	59	Yemen	84

Source: See endnote 7.

ers on the top 20 list in 2050—the Democratic Republic of the Congo, Uganda, and Yemen—are each projected to triple their populations by mid-century.[9]

What these demographic projections do not take into account are the constraints imposed by the capacity of life-support systems in individual countries. In many cases, the projection clearly exceeds the country's apparent ability to support its population. For example, the notion that Yemen—a country of 21 million people, where water tables are falling everywhere—will one day be able to support 84 million people requires a stretch of the imagination. Is Pakistan, with 158 million people today, likely to add nearly 200 million by 2050, making it larger than the United States today? And is it really possible that Nigeria will have 258 million people by 2050—almost as many as the United States has now?[10]

Population, Land, and Conflict

As land and water become scarce, we can expect mounting social tensions within societies, particularly between those who are poor and dispossessed and those who are wealthy, as well as among ethnic and religious groups, as competition for these vital resources intensifies. Population growth brings with it a steady shrinkage of life-supporting resources per person. That decline, which is threatening to drop the living standards of more and more people below survival level, could lead to unmanageable social tensions that will translate into broad-based conflicts.

Worldwide, the area in grain expanded from 590 million hectares (1,457 million acres) in 1950 to its historical peak of 730 million hectares in 1981. By 2004, it had fallen to 670 million hectares. Even as the world's population continues to grow, the area available for producing grain is shrinking.[11]

Expanding world population cut the grainland area per person in half, from 0.23 hectares (0.57 acres) in 1950 to 0.11 hectares in 2000. (See Figure 2–1.) This area of just over one tenth of a hectare per person is half the size of a building lot in an affluent U.S. suburb. This halving of grainland area per person makes it more difficult for the world's farmers to feed the 70 million or more people added each year. If current population projections materialize and if the overall grainland area remains constant, the area per person will shrink to 0.07 hectares in 2050, less than two thirds that in 2000.[12]

Having less cropland per person not only threatens livelihoods; in largely subsistence societies with nutrient-depleted soils, it threatens survival itself. Tensions among people begin to build as land holdings shrink below that needed for survival. The Sahelian zone of Africa, the broad swatch of the continent between the Sahara Desert and the more lush forested land to the south, which

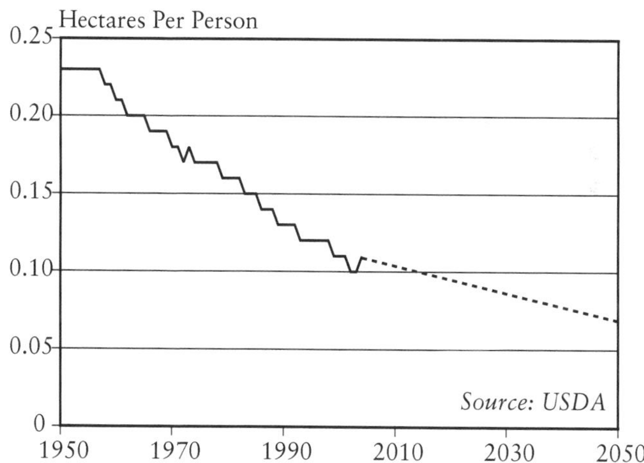

Figure 2–1. *World Grainland Per Person, 1950–2004, With Projection to 2050*

stretches from Sudan in the east through Mauritania in the west, has one of the world's fastest-growing populations. It is also an area of spreading conflicts.[13]

In troubled Sudan, 2 million people have died in the long-standing conflict between the Muslim north and the Christian south. The conflict in the Darfur region in western Sudan in 2004 illustrates the mounting tensions between two Muslim groups—Arab camel herders and black African subsistence farmers. Government troops are backing Arab militias, who are engaging in the wholesale slaughter of black Africans in an effort to drive them off their land, sending them into refugee camps in Chad.[14]

In Nigeria, where 130 million people are crammed into an area not much larger than Texas, overgrazing and overplowing are converting 351,000 hectares (1,350 square miles) of grassland and cropland into desert each year. The conflict between farmers and herders in Nigeria is a war for survival. As the *New York Times* reported in June 2004, "in recent years, as the desert has spread, trees have been felled and the populations of both herders and farmers have soared, the competition for land has only intensified."[15]

Unfortunately, the division between herders and farmers is also often the division between Muslims and Christians. This competition for land, amplified by religious differences and combined with a large number of frustrated young men with guns, has created what the *New York Times* describes as a "combustible mix" that has "fueled a recent orgy of violence across this fertile central Nigerian state [Kebbi]. Churches and mosques were razed. Neighbor turned against neighbor. Reprisal attacks spread until finally, in mid-May, the government imposed emergency rule."[16]

Similar divisions exist between herders and farmers in northern Mali, the *Times* noted, where "swords and sticks have been chucked for Kalashnikovs, as desertification and population growth have stiffened the competition between the largely black African farmers and the ethnic Tuareg and Fulani herders. Tempers are raw on both sides. The dispute, after all, is over livelihood and even more, about a way of life."[17]

Water, too, is a source of growing tension. Although much has been said about the conflicts between and among countries over water resources, some of the most bitter disagreements are taking place within countries where needs of local populations are outrunning the sustainable yield of wells. Local water riots are becoming increasingly common in countries like China and India. In the competition between cities and the countryside, cities invariably win, often depriving farmers of their irrigation water and thus their livelihood.[18]

The projected addition to the earth's population of 3 billion people by 2050, the vast majority of whom will be added in countries where water tables already are falling and wells are going dry, is not a recipe for economic progress and political stability. Continuing population growth in countries already overpumping their aquifers and draining their rivers dry could lead to acute hydrological poverty, a situation in which people simply do not have enough water to meet their basic needs.[19]

The Demographic Transition

In 1945, Princeton demographer Frank Notestein outlined a three-stage demographic model to illustrate the dynamics of population growth as societies modernized. (See Figure 2–2.) He pointed out that in pre-modern societies, births and deaths are both high and essentially in balance with little or no population growth. In stage two, as living standards rise and health care conditions improve, death rates begin to decline. With birth rates remaining high

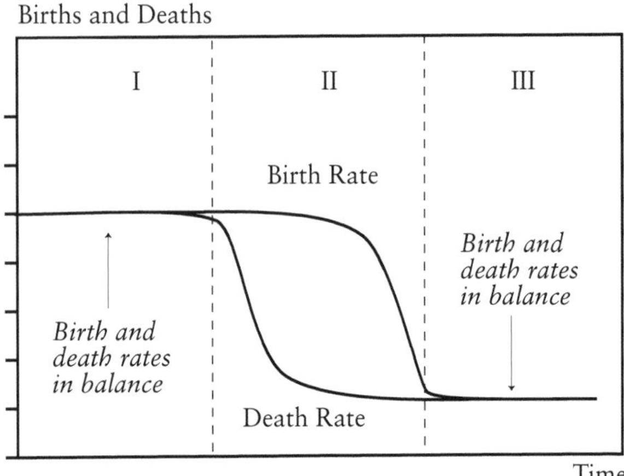

Figure 2–2. *The Three–Stage Process of the Demographic Transition*

while death rates are declining, population growth accelerates, typically reaching 3 percent a year. Although this may not sound like much, 3 percent a year results in a twenty-fold increase per century. As living standards continue to improve, and particularly as women are educated, the birth rate also begins to decline. Eventually the birth rate drops to the level of the death rate. This is stage three, where population is again stable.[20]

Of the 180 countries in the world today, some 36, with a combined population of 700 million people, have made it to stage three. With births and deaths essentially in balance, they have reached population stability. This leaves more than 140 countries—and 5.6 billion people—in stage two. Many with rising incomes and steadily declining birth rates are moving toward the population stability of stage three. Among them are China, Thailand, South Korea, and Iran. But many others in this group are not doing as well. After two generations of rapid growth, progress has largely come to a standstill. Living conditions in these largely rural societies are either improving very little or are deteriorating as family plots, divided and then subdivided, have left many families with too little land to sustain them.[21]

Stage two of the demographic transition, particularly the early part, is a politically risky place for countries to be. A study by Population Action International, *The Security Demographic: Population and Civil Conflict After the Cold War*, surveys the work of social analysts searching for advance indicators of political instability. One of the better known of these initiatives, a group known as the State Failure Task Force and set up by the Central Intelligence Agency in the 1990s, tried to determine what social, political, economic, and environmental variables could help anticipate what they termed "state failure." This, in effect, is a form of social disintegration, a collapse of order in a society. Of all the indicators analyzed by the task force, high infant mortality correlated most closely with political instability.[22]

The second best indicator of political volatility was a disproportionately large share of the population in the young adult category, those in their late teens and twenties. The prospect that large numbers of young adults would foster social conflict and political instability was much stronger in societies where educational and economic opportunities were lacking.[23]

Once countries have moved into the final stage of the demographic transition, when both mortality and fertility are low and essentially in balance, the chance of civil conflicts diminishes sharply. This suggests that it is in the global interest to help those countries that are stalled in stage two to get moving and make it into stage three as soon as possible.

The progression through stage two of the demographic transition is not a smooth one and it is by no means automatic. While there is no evidence of a country that has made it to stage three falling back into stage two, there is growing evidence that countries remaining in stage two for an inordinate amount of time are falling back into stage one.[24]

Governments in countries that have experienced rapid population growth for nearly two generations are showing signs of "demographic fatigue." Worn down by the struggle to feed, clothe, educate, and provide health care for an ever-expanding population, they are unable to respond to new threats, such as HIV/AIDS.[25]

Countries that remain in stage two, with its rapid population growth, risk being overwhelmed by land hunger, water shortages, disease, civil conflict, and other adverse effects of prolonged rapid population growth. Yemen, Ethiopia, the Democratic Republic of the Congo, Somalia, and Afghanistan all fall into this category. Among the countries that are sliding back into stage one—where high death rates offset high birth rates, thus preventing any population growth—are Botswana and South Africa.[26]

Within the next two decades or so, most of the countries in stage two will either have made it into stage three or fallen back to stage one. What is not clear is exactly what combination of events and forces will push countries backward demographically. At this point, it is obvious that the HIV epidemic is responsible for the handful of countries that are moving back toward stage one, where rising mortality may not merely balance fertility but exceed it, leading to an absolute decline in population. Countries where a fifth or more of adults are HIV-positive will lose a comparable share of their adult populations within the next decade or so. For each adult sick with AIDS, another adult typically provides care. As the virus spreads, the number of people able to till the fields shrinks, until eventually food production falls. At this point, disease and hunger reinforce each other in a downward spiral leading countries into a demographic dark hole.[27]

The Demographic Bonus

In contrast to these countries whose future is fading, countries that have quickly reduced birth rates are benefiting from what economic demographers have labeled a "demographic bonus." When a country shifts quickly to smaller families, the number of young dependents—those who need nurturing and educating—declines sharply relative to the number of working adults. In this situation, household savings climb, investment rises, worker productivity increases, and economic growth accelerates. Since European countries did not experience the rapid population growth of today's developing countries, and therefore no rapid fall in fertility, they never experienced a demographic bonus.[28]

Virtually all countries that have quickly shifted to smaller families have benefited from the demographic bonus. When Japan cut its population growth rate in half between 1951 and 1958, for instance, it became the first country to benefit from this bonus. The spectacular economic growth in the 1960s, 1970s, and 1980s, unprecedented in any country, raised Japan's income per person to one of the highest in the world, making it a modern industrial economy second in size only to the United States.[29]

South Korea, Taiwan, Hong Kong, and Singapore followed shortly thereafter. These four so-called tiger economies, which enjoyed such spectacular economic growth during the late twentieth century, each benefited from a rapid fall in birth rates and the demographic bonus that followed.[30]

On a much larger scale, China's sharp reduction in its birth rate created a large demographic bonus and a population that saves more than 30 percent of its income for investment. This phenomenal investment rate, coupled with the record influx of private foreign investment and accompanying technology, is propelling China into the ranks of modern industrial powers.[31]

China is the most highly visible of a second wave of countries that are likely to benefit from the demographic bonus. The Population Action International study indicates that other countries with age structures now favorable to high savings and rapid economic growth include Sri Lanka, Mexico, Iran, Tunisia, and Viet Nam.[32]

After a point, growth in the labor force begins to slow as the results of the falling birth rate are reflected in the shrinking number of entrants into the labor force. This in turn leads to higher wages. Women respond to these by entering the work force, which contributes to a further decline in fertility—one that in some countries is leading to an actual decline in population size.[33]

Two Success Stories

Some countries with fast-growing populations that face fast-shrinking water and cropland availabilities per person fail to slow their population growth and, as a result, experience spreading hunger and political instability. Other countries see the handwriting on the wall and move to quickly slow their population growth.

The good news is that countries that want to reduce family size quickly can do so. Two of the best examples of this are Thailand and Iran. These two middle-sized countries have been remarkably successful in slowing population growth, although they have very different cultures and economies. While Thailand's farm economy is rice-based, Iran's is wheat-based. Thailand is humid and subtropical, while Iran is semiarid and temperate. One nation is predominantly Buddhist, the other Muslim.[34]

Thailand's success can largely be traced to one individual, Mechai Viravaidya, who eventually became known nationwide simply as Mechai. During the 1970s Mechai saw that if Thailand did not rein in its population growth, it would eventually be in serious trouble. He recognized early on that family planning, reproductive health, and contraception were topics that people needed to feel comfortable talking about.[35]

One of his first goals was thus to promote the discussion of population and family planning issues. He gave talks to any group who would listen. He worked with educators to get population examples in elementary school math books. He wanted even Thailand's children to understand the consequences of prolonged exponential growth.[36]

He popularized the condom, one of the first contraceptives available in Thailand, and promoted its manufacture and distribution. He helped people understand the role of condoms in preventing births and disease. Schoolchildren played games with condoms inflated as balloons. Taxi drivers in Bangkok had condoms in their cabs, offering them to their passengers for free. At a 1979 conference of Parliamentarians on Population and Development that I attended in Colombo, Sri Lanka, Mechai boarded a bus to the meeting site and went down the aisle with a small box filled with condoms, offering them to various members of Parliament—men and women alike—teasing them about the colors they wanted or the size that would be best for them. He was thoroughly entertaining—and certainly disarming—which is no doubt why "Mechai" is now slang for condom in Thailand.[37]

Mechai's enthusiasm could not be curbed. The bottom line was that he mobilized the resources of the Thai government to introduce family planning programs throughout the country. In 2000, Mechai was elected to the Senate by the people of Thailand.[38]

Today, women in Thailand have access to a full range of family planning services. Instead of a population growth rate of 3 percent a year—or twentyfold per century—Thailand's annual population growth rate is 0.8 percent. With the average number of children per woman in Thailand now less than 2, it is only a matter of time until Thailand's population stabilizes. Its current population of 63 million is projected to stop growing at around 77 million by 2050, an increase of 22 percent. This compares with the projected growth of 38 percent for the United States by 2050.[39]

Iran's dramatic gains in reducing family size have come more recently. In scarcely a decade, Iran reduced its population growth from the world's highest of nearly 4 percent a year to just over 1 percent. The country's roller-coaster population policy began when Ayatollah Khomeini replaced the Shah in 1979. One of the first things Khomeini did was to dismantle the family planning programs the Shah had introduced in 1967. Khomeini then advocated large families. Between 1980 and 1988, Iran was continuously at war with Iraq, and Khomeini wanted large families to produce more soldiers. His aim was eventually to field an army of 20 million troops. As women were urged to have more and more children, the population growth rate hit 4.4 percent in the early 1980s, close to the biological maximum and one of the highest ever recorded.[40]

A decade later, Iran reversed its population policy by 180 degrees. The country's leadership had crossed a threshold, recognizing that their record population growth was burdening the economy, destroying the environment, and overwhelming schools. They then started a family planning program to reduce family size.[41]

Overnight they launched a new program that quickly became one of the most comprehensive efforts to slow population growth ever adopted in any country. This program was not left to the family planners alone. The government also mobilized the ministries of education and culture to help convince the public of the need to shift to smaller families and to slow population growth.[42]

Iran Broadcasting played a prominent role, releasing a steady drumbeat of information encouraging smaller families and extolling their benefits. Radio and television broadcasts informed people that family planning services were available. Indeed, it let them know of the 15,000 new "health houses" available in villages to provide family planning guidance and services. The national female literacy rate climbed from roughly 25 percent in 1970 to over 70 percent today.[43]

Religious leaders were mobilized to convince couples to have smaller families. Mullahs who once were on the front lines urging women to have more children were now encouraging them to have fewer. Iran pioneered with a family planning program that offered the entire range of contraceptive practices and materials. Contraceptives, such as the pill, were free of charge. Iran also became the first Muslim country to offer male sterilization. And uniquely, in Iran couples must take a two-day course in family planning and contraception in order to get a marriage license.[44]

Average family size has dropped from seven children to fewer than three. The population growth rate was cut in half from 1987 to 1994, putting Iran in the same category as Japan and China—the only other two countries that have succeeded in halving their population growth rates in such a short period of time. In 2004, Iran's population was growing only modestly faster than that of the United States.[45]

If Iran, with its strong undercurrent of Islamic fundamentalism, can move so quickly toward population stability, then there is hope for countries everywhere. Over the long term a sustainable population means two children per couple. The arithmetic is simple. Any population that increases or decreases continuously over the long term is not sustainable.

Eradicating Poverty, Stabilizing Population

Stabilizing population is the key to maintaining political stability and sustaining economic progress. And the keys to stabilizing population are universal elementary-school education, basic health care, access to family planning, and, for the poorest of the poor countries, school lunch programs.

The United Nations has established universal primary school education by 2015 as one of its Millennium Development Goals. This means educating all children, but with a special focus on girls, whose schooling has lagged behind that of boys in almost all developing countries. The more education girls get, the fewer children they have. This is a relationship that cuts across all cultures and societies. As educational levels go up, fertility levels come down.[46]

Closely related to universal primary school education is basic health care, village-level care of the most rudimentary kind. It includes rural clinics that provide childhood immunization for infectious diseases, oral rehydration therapy to cope with dysentery, reproductive health care, and family planning services along the lines of Iran's rural "health houses." In the poorest of the poor countries, where infant mortality rates are still high, parents remain reluctant to have fewer children because there is so much uncertainty about how many will survive to adulthood to look after them.[47]

School lunch programs are needed in poor countries for two reasons. One, they provide an incentive for poor children, often weakened by hunger, to make it to school. Two, once children are in school, having food helps them learn. If children are chronically hungry, their attention spans are short.[48]

We all have a stake in ensuring that countries everywhere move into stage three of the demographic transition. Countries that fall back into stage one are likely to be politically unstable—ridden with ethnic, racial, and religious conflict. These failed states are more likely to be breeding grounds for terrorists than participants in building a stable world order.

If world population continues to grow at 70 million or more people per year, the number of people trapped in hydrological poverty and hunger will almost certainly grow, threatening food security, political stability, and economic progress. The only humane option is to move quickly to a two-child family and try to stabilize world population at closer to 7 billion than the 9 billion currently projected. Against this backdrop, the time has come for world leaders, including the Secretary-General of the United Nations, the President of the World Bank, and the President of the United States, to recognize publicly that the earth cannot support more than two children per family over the long term.

Data for figures and additional information can be found at www.earth-policy.org/Books/Out/index.htm.

3

Moving Up the Food Chain Efficiently

Throughout most of our 4 million years as a distinct species we lived as hunter-gatherers. The share of our diets that came from hunting or gathering varied with geographic location, skills, and the season of the year. During the northern hemisphere winter, when there was little to gather, we depended heavily on hunting for our survival. This long history as hunter-gatherers left us with an appetite for animal protein, one that continues to shape diets today.

In every country where incomes have risen, this appetite for meat, eggs, and seafood has generated an enormous growth in animal protein consumption. The form the animal protein takes depends heavily on geography. Countries that are land-rich with vast grasslands depend heavily on beef—the United States, Brazil, Argentina, Australia, and Russia—or on mutton, as in Australia and Kazakhstan. Countries that are more densely populated and lack extensive grazing lands have historically relied much more on pork. Among these are Germany, Poland, and China. Densely populated countries with long shorelines, such as Japan and Norway, have turned to the oceans for their animal protein.[1]

While we typically focus on the food requirements

generated by population growth and the pressure this puts on the earth's land and water resources, moving up the food chain also adds to the pressure. The challenge is to do so as efficiently as possible, minimizing additional demands on land and water. Encouragingly, new approaches to the production of livestock, poultry, and fish are raising the efficiency with which grain is converted into animal protein.

Up the Food Chain

For those living at subsistence level, 60 percent or more of calories typically come from a single starchy food staple such as rice, wheat, or corn. Diversifying this diet is everywhere a high personal priority as incomes rise. One of the first additions people make is animal protein in some form—meat, milk, eggs, and fish.[2]

Since 1950, world meat production has climbed from 44 million to 253 million tons, more than a fivefold jump. Except for 1959, it has risen every year during this period, becoming one of the world's most predictable economic trends. (See Figure 3–1.) Worldwide, the average person consumed 41 kilograms of meat in 2003, more than double the figure a half-century ago.[3]

Comparing grain use per person in India and the United States gives us an idea of how much grain it takes to move up the food chain. In a low-income country like India—where annual grain production falls well short of 200 kilograms per person, or roughly 1 pound a day—nearly all grain must be eaten directly to satisfy basic food energy needs. Little can be converted into animal protein. Not surprisingly, the consumption of most livestock products in India, especially meat where religious restrictions also apply, is rather low. Milk, egg, and poultry consumption, however, are beginning to rise, particularly among India's expanding middle class.[4]

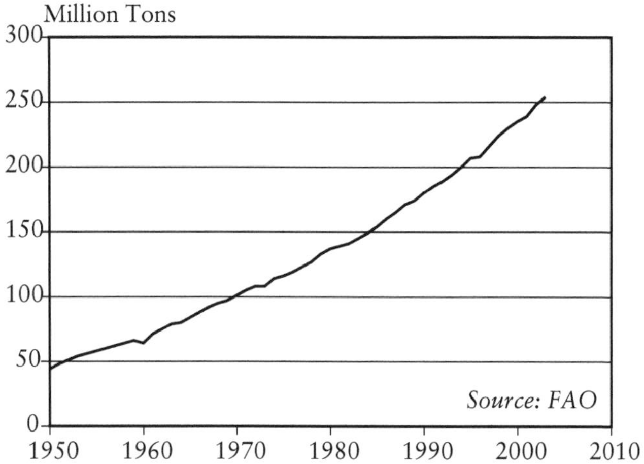

Figure 3–1. *World Meat Production, 1950–2003*

The average American, in contrast, consumes roughly 800 kilograms of grain per year, four fifths or more of it indirectly in the form of meat, milk, eggs, and farmed fish. Thus the grain consumption, direct and indirect, of an affluent American is easily four times that of the typical Indian.[5]

Ironically, the healthiest people in the world are not those who live very low or very high on the food chain but those who occupy an intermediate position. Italians, eating less than 400 kilograms of grain per person annually, have a longer life expectancy than either Indians or Americans. This is all the more remarkable because U.S. expenditures on health care per person are much higher than those in Italy. Italians benefit from what is commonly described as the Mediterranean diet, considered by many to be the world's healthiest.[6]

People in some countries live high on the food chain but use relatively little grain to feed animals; Argentina and Brazil, for instance, depend heavily on grass-fed beef.

Japanese also live high on the food chain, but use only moderate amounts of feedgrains because their protein intake is dominated by seafood from oceanic fisheries.[7]

Shifting Protein Sources

The composition of world meat production has changed dramatically over the last half-century or so. From 1950 until 1978, beef and pork vied for the lead. (See Figure 3–2.) Then the world meat consumption pattern began to change as economic reforms adopted in China in 1978 led to a dramatic climb in pork production, pushing it far ahead of beef worldwide.[8]

In an effort to minimize waste, village families in China have a long-standing tradition of keeping a pig, which is fed all the kitchen and table waste. When the pig matures, it is butchered and eaten and replaced with another small, recently weaned, pig. Even today, four fifths of China's pork production takes place at the family level.[9]

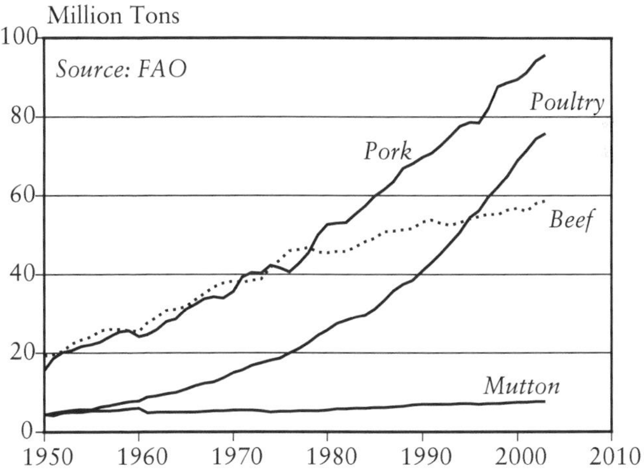

Figure 3–2. *World Meat Production by Type, 1950–2003*

With China's 1.3 billion people clamoring for more pork, production there climbed from 9 million tons in 1978, the year of the economic reforms, to 46 million tons in 2003. U.S. pork production rose only from 6 million to 9 million tons during the same period, and pork consumption per person in China overtook that in the United States. Perhaps even more impressive, half of the world's pork is now eaten in China.[10]

In 1950, when beef and pork dominated world meat consumption, poultry production was quite low, roughly the same as mutton. From mid-century onward, however, poultry production gathered momentum, overtaking beef in 1996. Advances in the efficiency of poultry production had dropped the price to the point where more and more people could afford it. In the United States—where a half-century ago chicken was something special, usually served only at Sunday dinner—its low price now makes chicken the meat of choice for everyday consumption.[11]

With overgrazing widespread, additional beef production now comes largely from putting more cattle in feedlots for a longer period of time. Thus the changing composition of our diets reflects the widely varying efficiency with which cattle, pigs, chickens, and, increasingly, fish convert grain into protein. A steer in a feedlot requires 7 kilograms of grain for each kilogram of weight gain. For pork, each kilogram of additional live weight requires about 3.5 kilograms. For poultry, it is just over 2. For catfish in the United States and carp in China and India, it is 1–2 kilograms of feed per kilogram of additional weight gain.[12]

Between 1990 and 2003, growth in beef production averaged less than 1 percent a year. Pork, meanwhile, expanded at 2.5 percent annually, eggs at nearly 4 percent, and poultry at 5 percent. Aquacultural output, which sets the gold standard in the efficiency of feed conversion into protein, expanded by nearly 10 percent a year, climbing from 13 million tons in 1990 to 40 million tons in 2002. (See Table 3–1.)[13]

Historically, as the demand for seafood increased with rising incomes, countries turned to the oceans. As population pressure built up, for example, beginning a century or so ago, Japan needed nearly all its arable land to produce rice, leaving almost none for producing feed for livestock and poultry. So the country started relying more on fish for animal protein and now consumes 10 million tons of seafood per year. But with oceanic fisheries being pushed to their limits, there are few opportunities for countries developing appetites for animal protein to switch to eating fish in the same way. For example, if China's per capita consumption of seafood from oceanic fisheries reached the Japanese level, the country would need 100 million tons of seafood—more than the world catch.[14]

Table 3–1. *Annual Growth in World Animal Protein Production, by Source, 1990–2003*

Source	1990	2003	Annual Growth
	(million tons)		(percent)
Beef	53	59	0.8
Pork	70	96	2.5
Mutton	10	12	1.6
Poultry	41	76	4.9
Eggs	38	61	3.7
Oceanic Fish Catch	85	93[1]	0.8
Aquacultural Output	13	40[1]	9.7

[1]Figures for 2002.

Source: See endnote 13.

So although China is a leading claimant on oceanic fisheries, with a catch of 16 million tons per year, it has turned to fish farming to satisfy most of its fast-growing seafood needs and is leading the world into the aquaculture era. China's aquacultural output, mainly carp and shellfish, totals 28 million tons. With incomes now rising in densely populated Asia, other countries are following China's lead. Among them are India, Thailand, and Viet Nam. Viet Nam, for example, devised a plan in 2001 of developing 700,000 hectares of land in the Mekong Delta for aquaculture, with the goal of producing 1.7 million tons of fish and shrimp by 2005. It now appears likely to exceed this goal.[15]

Over the last 15 years, aquaculture has thus emerged as a major source of animal protein. Driven by the high efficiency with which omnivorous species, such as carp, tilapia, and catfish, convert grain into animal protein, world aquacultural output nearly tripled between 1990 and 2002. It will likely overtake beef production worldwide by 2010.[16]

As the consumption of animal protein has grown, the share of the world grain harvest used for feed has remained constant at roughly 37 percent for two decades. Of the world's three leading grains—rice, wheat, and corn—which together account for nearly 90 percent of the grain harvest, rice is grown almost entirely as a food crop. Wheat is largely a food crop, but one sixth of the wheat harvest is fed to livestock and poultry. In contrast, the world's huge corn harvest is consumed largely as feed. In recent years, the addition of a protein supplement (typically soybean meal) to feed rations has boosted the efficiency of feed conversion into animal protein. This stabilized the share of the world grain harvest used for feed even while meat, milk, and egg consumption per person were climbing.[17]

Oceans and Rangelands

During much of the last half-century, the growth in demand for animal protein was satisfied by the rising output of two natural systems: oceanic fisheries and rangelands. Between 1950 and 1990, the oceanic fish catch climbed from 19 million to 85 million tons, a five-fold gain. (See Figure 3–3.) During this period, seafood consumption per person nearly doubled, climbing from 8 to 15 kilograms.[18]

This was the golden age of oceanic fisheries. Never before had the world seen such growth in an animal protein source. It grew rapidly as fishing technologies evolved that helped fishers bring in their catch more efficiently and as refrigerated processing ships began to accompany fishing fleets, enabling them to operate in distant waters.

Unfortunately, the human appetite for seafood is outgrowing the sustainable yield of oceanic fisheries. Today

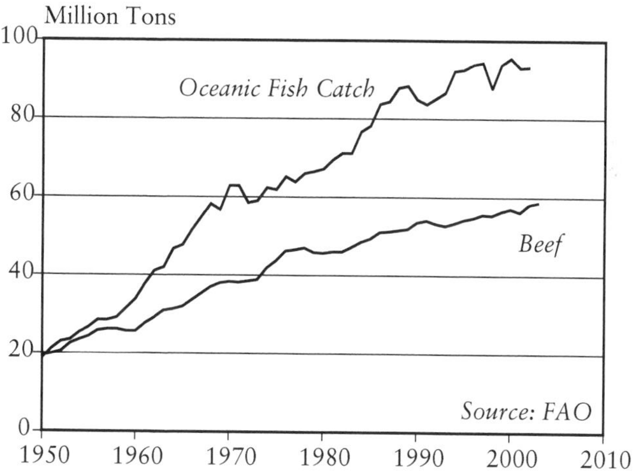

Figure 3–3. *World Oceanic Fish Catch and Beef Production, 1950–2003*

70 percent of fisheries are being fished at or beyond their sustainable capacity. As a result, many are in decline and some have collapsed. In some fisheries, the breeding stocks have been mostly destroyed. A 2003 landmark study by a Canadian-German science team, published in *Nature*, concluded that 90 percent of the large predatory fish in the oceans had disappeared over the last 50 years.[19]

This ambitious, 10-year assessment drew on data from all the world's major fisheries. Ransom Myers, a fisheries biologist at Canada's Dalhousie University and lead scientist in this study, says, "From giant blue marlin to mighty blue fin tuna, from tropical groupers to Antarctic cod, industrial fishing has scoured the global ocean. There is no blue frontier left."[20]

Fisheries are collapsing throughout the world. The fabled cod fishery of Canada failed in the early 1990s. Those off the coast of New England were not far behind. And in Europe, cod fisheries are in decline, approaching a free fall. Like the Canadian cod fishery, the European fisheries may have been depleted to the point of no return.[21]

Myers goes on to say, "Since 1950, with the onset of industrialized fisheries, we have rapidly reduced the resource base to less than 10 percent—not just in some areas, not just for some stocks, but for entire communities of these large fish species from the tropics to the poles." In contrast to the rapid rise in seafood consumption per person during the last century, the world's still-growing population is now facing a substantial decline in seafood catch per person.[22]

Rangelands, like the oceans, are also essentially natural systems. Located mostly in semiarid regions too dry to sustain agriculture, they are vast, covering roughly twice the area planted to crops.[23]

Perhaps 180 million of the world's people depend

entirely on livestock for their livelihood. Most of these are in the pastoral communities of Africa, the Middle East, Central Asia, Mongolia, and northern and western China. As these populations grew, so did their livestock populations. Environmentally, nearly all these pastoral societies are in trouble. Since rangelands are typically owned in common, there is no immediate reason for individual families to limit the number of cattle, sheep, or goats. The result is widespread overgrazing, desertification, personal hardship, and slower growth in livestock production.[24]

World beef production climbed from 20 million tons in 1950 to nearly 60 million tons in 2003. But after doubling between 1950 and 1975, growth has become progressively slower, dropping to under 1 percent a year since 1990. Just as oceanic fisheries are being overfished, the world's rangelands are being overgrazed. As a result, the grasses on which livestock forage are slowly deteriorating.[25]

As vegetation disappears, the soil begins to blow. At first dust storms remove the finer particles of soil. Once these have largely blown away, sand storms become the prevailing measure of degradation. As the sand begins to drift, it forms sand dunes; these begin to encroach on farmers' land, making both grazing and farming untenable.

The world has reached the end of an era with both oceanic fisheries and rangelands. Human demands for seafood, beef, and mutton have surpassed the sustainable yield of these systems. With these two natural systems reaching their limits, future growth in animal protein production will have to come largely from feeding. Producing the feedstuffs, principally corn and soybeans, will put more pressure on the earth's land and water resources—pressure that is already unsustainable in some countries. At this point, the incorporation of soybean meal into livestock rations to boost sharply the efficiency with which grain is converted into animal protein is indispensable.

The Soybean Factor

When we think of soybeans in our daily diet, it is typically as tofu, veggie burgers, or other meat substitutes. But most of the world's fast-growing soybean harvest is consumed indirectly in the beef, pork, poultry, milk, eggs, and farmed fish that we eat. Although not a visible part of our diets, the incorporation of soybean meal into feed rations has revolutionized the world feed industry, greatly increasing the efficiency with which grain is converted into animal protein.[26]

In 2004, the world's farmers produced 223 million tons of soybeans, 1 ton for every 9 tons of grain they produced. Of this, some 15 million tons were consumed as tofu or meat substitutes. The remaining 208 million tons were crushed in order to extract 33 million tons of soybean oil, separating it from the more highly valued meal. Soybean oil dominates the world vegetable oil economy, supplying much of the oil used for cooking and to dress salads. Soybean oil production exceeds that of all the other table oils combined—olive, safflower, canola, sunflower, and palm oil.[27]

The 143 million tons of soybean meal that remains after the oil is extracted is fed to cattle, pigs, chicken, and fish, enriching their diets with high-quality protein. Experience in feeding shows that combining soybean meal with grain, in roughly one part meal to four parts grain, dramatically boosts the efficiency with which grain is converted into animal protein, sometimes nearly doubling it.[28]

The world's three largest meat producers—China, the United States, and Brazil—now all rely heavily on soybean meal as a protein supplement in feed rations. The United States has long used soybean meal to upgrade livestock and poultry feed. As early as 1964, 8 percent of feed rations consisted of soybean meal. Over most of the last decade, the meal content of U.S. feeds has fluctuated between 17 and 19 percent.[29]

For Brazil, the shift to soybean meal as a protein supplement began in the late 1980s. From 1986 to 1997, the soymeal share of feed rations jumped from 2 percent to 21 percent. In China, the realization that feed use efficiency could be dramatically boosted with soymeal was translated into reality some six years later. Between 1991 and 2002, the soymeal component of feed jumped from 2 percent to 20 percent. For fish, whose protein demands are particularly high, China incorporated some 5 million tons of soymeal into the 16 million tons of grain-based fish feed used in 2003.[30]

The experience of these three countries simply indicates that the same principles of animal nutrition apply everywhere. The ratio of soybean meal to corn in the feed mix varies somewhat according to the price relationship between the two. Where corn is cheap, as in the United States, the corn share of the feed mix tends to be slightly higher. In Brazil, which has an economic advantage in soybean production, the soy component is higher.[31]

As world grain production was tripling from 1950 to 2004, soybean production was expanding thirteenfold. The growth in this protein source, most of it consumed indirectly in various animal products, is a surrogate for rising affluence, one that measures movement up the food chain.[32]

The soybean was domesticated in central China some 5,000 years ago and made its way to the United States in 1804, when Thomas Jefferson was President. For a century and a half the soybean was grown mostly as a curiosity crop in home gardens. Most farmers outside of China did not even know what a soybean looked like. But after World War II, production exploded as the consumption of livestock and poultry products climbed in North America and Europe.[33]

By 1978, the area planted to soybeans in the United States had eclipsed that planted to wheat. In some recent years, the U.S. harvested area of soybeans has exceeded that of corn, making it the country's most widely planted crop. In the United States, where soybean production is now five times that in China, the soybean has found an ecological and economic niche far larger than in its country of origin.[34]

U.S. soybeans are grown mostly in the Corn Belt, often in rotation with corn. The soybean, a nitrogen-fixing legume, and corn, which has a ravenous appetite for nitrogen, fit together nicely on the same piece of land in alternate years. In fact, if the Corn Belt were being named today, it would be called the Corn/Soybean Belt.[35]

Another chapter in the soybean saga has been unfolding over the past three decades in Latin America. After the collapse in 1972 of the Peruvian anchovy fishery—which accounted for a fifth of the world fish catch and supplied much of the protein meal used in livestock and poultry foods at that time—some countries in Latin America saw an opportunity to produce soybeans. As a result, both Brazil and Argentina began to expand soybean production, slowly at first and then, during the 1990s, at breakneck speed. As of 2004, soybean production exceeds that of all grains combined in both countries. Brazil now exports more soybeans than the United States does. And within the next few years Brazil is likely to overtake the United States in production as well.[36]

While production was increasing thirteenfold over the last half-century, soybean yields have almost tripled, which means that the area in soybeans has increased some fourfold. In contrast to grains, where the growth in output has come largely from raising yields, growth in the harvest of the land-hungry soybean has come more from area expansion.[37]

As a result, in a world with limited cropland resources, the soybean has been expanding partly at the expense of grain. Nonetheless, this expansion so greatly increases the efficiency of grain used for feed that it reduces the cropland area used to produce feedgrains and soybeans together.[38]

New Protein Models

Mounting pressure on the earth's land and water resources to produce livestock, poultry, and fish feed has led to the evolution of some promising new animal protein production models, one of which is milk production in India. Since 1970, India's milk production has increased more than fourfold, jumping from 21 million to 87 million tons. In 1997, India overtook the United States in dairy production, making it the world's leading producer of milk and other dairy products. (See Figure 3–4.)[39]

The spark for this explosive growth came in 1965 when an enterprising young Indian, Dr. Vargese Kurien, organized the National Dairy Development Board, an umbrella organization of dairy cooperatives. A coop's principal purpose was to market the milk from tiny herds that typically averaged two to three cows each. It was these dairy cooperatives that provided the link between the growing appetite for dairy products and the millions of village families who had only a few cows and a small marketable surplus.[40]

Creating the market for milk spurred the fourfold growth in output. In a country where protein shortages stunt the growth of so many children, expanding the milk supply from less than half a cup per person a day 25 years ago to more than a cup represents a major advance.[41]

What is new here is that India has built the world's largest dairy industry almost entirely on roughage—wheat straw, rice straw, corn stalks, and grass collected

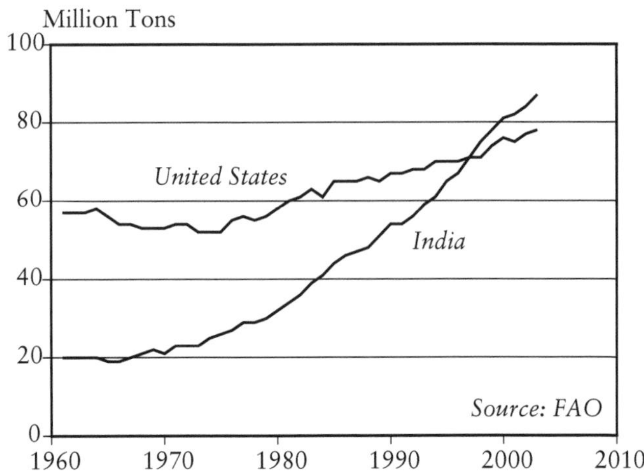

Figure 3–4. *Milk Production in India and the United States, 1961–2003*

from the roadside. Cows are often stall-fed with crop residues or grass gathered daily and brought to them.[42]

A second new protein production model, which also relies on ruminants, is one that has evolved in China, principally in four provinces of central Eastern China—Hebei, Shangdong, Henan, and Anhui—where double cropping of winter wheat and corn is common. Once the winter wheat matures and ripens in early summer, it must be harvested quickly and the seedbed prepared to plant the corn. The straw that is removed from the land in preparing the seedbed as well as the cornstalks left after the corn harvest in late fall are fed to cattle. Although these crop residues are often used by the villagers as fuel for cooking, they are shifting to other sources of energy for cooking, which lets them keep the straw and cornstalks for feed. By supplementing this roughage with small amounts of nitrogen, typically in the form of urea, the microflora in the complex four-stomach digestive system of cattle can convert roughage efficiently into animal protein.[43]

This practice enables these four crop-producing provinces to produce much more beef than the vast grazing provinces in the northwest do. This central eastern region of China, dubbed the Beef Belt by Chinese officials, is producing large quantities of animal protein using only roughage. The use of crop residues to produce milk in India and beef in China means farmers are reaping a second harvest from the original crop.[44]

Another promising new animal protein production model has also evolved in China, this one in the aquacultural sector. China has evolved a carp polyculture production system in which four species of carp are grown together. One species feeds on phytoplankton. One feeds on zooplankton. A third feeds on grass. And the fourth is a bottom feeder. These four species thus form a small ecosystem, with each filling a particular niche. This multi-species system, which converts feed into flesh with remarkable efficiency, yielded some 13 million tons of carp in 2002.[45]

While poultry production has grown rapidly in China over the last two decades, it has been dwarfed by the phenomenal growth of aquaculture. (See Figure 3–5.) Today aquacultural output in China—at 28 million tons—is double that of poultry, making it the first country where aquaculture has emerged as a leading source of animal protein. The great economic and environmental attraction of this system is the efficiency with which it produces animal protein.[46]

Although these three new protein models have evolved in India and China, both densely populated nations, they may find a place in other parts of the world as population pressures intensify and as people seek new ways to convert plant products into animal protein.

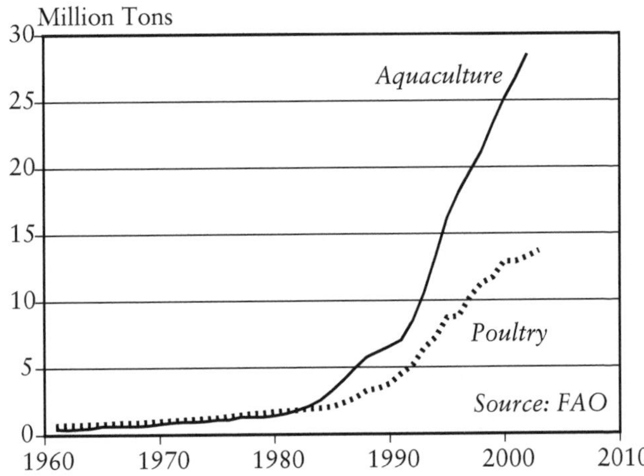

Figure 3–5. *Aquacultural and Poultry Production in China, 1961–2003*

The world desperately needs more new protein production techniques such as these. A half-century ago, when there were only 2.5 billion people in the world, virtually everyone wanted to move up the food chain. Today there may be close to 5 billion people wanting more animal protein in their diet. The overall demand for meat is growing at twice the rate of population; the demand for eggs is growing nearly three times as fast; and growth in the demand for fish—both from the oceans and from fish farms—is also outpacing that of population. Against this backdrop of growing world demand, our ingenuity in producing animal protein in ever-larger quantities and ever more efficiently is going to be challenged to the utmost.[47]

While the world has had many years of experience in feeding an additional 70 million or more people each year, it has no experience with some 5 billion people wanting to move up the food chain at the same time. For a sense of what this translates into, consider what has happened in China since the economic reforms in 1978. As the fastest-growing economy in the world since 1980, China has in effect telescoped history, showing how diets change when incomes rise rapidly over an extended period.[48]

As recently as 1978, meat consumption was low in China, consisting mostly of modest amounts of pork. Since then, consumption of pork, beef, poultry, and mutton has climbed. In 2003 people in China ate some 71 million tons of meat, close to twice as much as Americans ate. China has decisively displaced the United States, long number one in meat consumption. (See Figure 3–6.)[49]

As incomes rise in other developing countries, people will also want to increase their consumption of animal protein. Considering the demand this will place on the

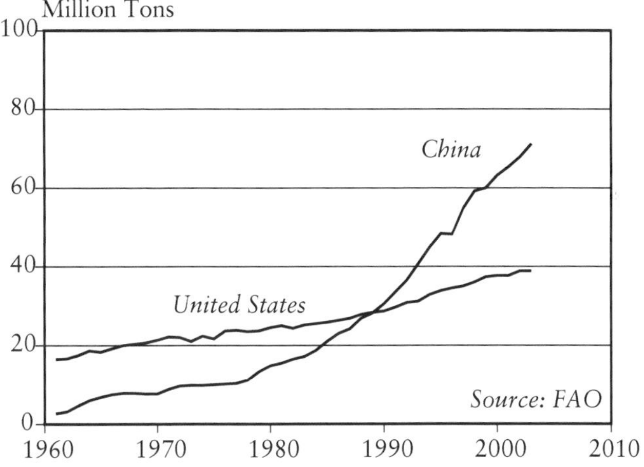

Figure 3–6. *Meat Production in China and the United States, 1961–2003*

earth's land and water resources, along with the more traditional demand from population growth, provides a better sense of the future pressures on the earth. If world grain supplies tighten in the years ahead, the competition for this basic resource between those living high on the food chain and those on the bottom rungs of the economic ladder will become both more visible and a possible source of tension within and among societies.

Data for figures and additional information can be found at www.earth-policy.org/Books/Out/index.htm.

4

Raising the Earth's Productivity

During the last half of the twentieth century the world's farmers more than doubled the productivity of their land, raising grain yield per hectare from 1.1 tons in 1950 to 2.7 tons in 2000. Never before had there been an advance remotely approaching this one. And there may not be another.[1]

The unprecedented gains in land productivity were the result of the systematic application of science to agriculture. The early gains were based primarily on research by governments in Japan, the United States, and Europe. In the United States, the U.S. Department of Agriculture (USDA) orchestrated the national effort while Agricultural Experiment Stations located at land-grant universities in each state focused on the specific research needs of local farmers. Then as agriculture advanced, agribusiness firms producing seed, fertilizer, pesticides, and farm equipment invested heavily in the development of technologies that would help expand food production. Today the lion's share of agricultural research is funded by corporations.[2]

The strategy of systematically applying science to agriculture while simultaneously providing economic incentives to farmers to expand output was phenomenally successful. Between 1950 and 1976, the annual world grain harvest doubled, going from 630 million to 1,340 million tons. In a single generation, the world's farmers expanded grain production by as much as they had during the preceding 11,000 years since agriculture began.[3]

Trends and Contrasts

Throughout most of history, rises in farmland productivity were so slow as to be imperceptible within a given generation. When Japan succeeded in launching a sustained rise in rice yields in the 1880s, it became the first country to achieve a "takeoff" in grain yield per hectare. But it was not until World War II that other industrial countries, including the United States and the countries of Europe, also initiated steady rises in cropland productivity.[4]

Plant breeding programs in Japan that gave the world the dwarf rices and wheats and the U.S. programs that yielded hybrid corn were at the center of the revolutionary rises in land productivity. By the mid-1960s, developing countries such as India were also beginning to raise yields. Using a combination of price incentives and a modified version of Japan's high-yielding dwarf wheats that were developed at the International Maize and Wheat Improvement Center in Mexico, India doubled its wheat crop between 1965 and 1972. This was the fastest doubling of a grain harvest in a major country on record. Other countries, including Pakistan and Turkey, also moved quickly to raise wheat yields, although China's big jump in grain yields did not come until after the economic reforms of 1978.[5]

Early success in adapting the high-yielding dwarf wheats in Mexico led to an intense effort to adapt Japan's dwarf rices to tropical and subtropical growing conditions throughout Asia. Indeed, the International Rice Research Institute (IRRI) in the Philippines was founded in 1960 by the Rockefeller and Ford Foundations specifically for this purpose.[6]

The record rise in world grainland productivity since 1950 had three sources—genetic advances, agronomic improvements, and synergies between the two. The genetic contribution to raising yields has come largely from increasing the share of the plant's photosynthetic product (the photosynthate) going to seed. Shifting as much photosynthate as possible from the leaves, stems, and roots to the seed helps to maximize yields. For example, the originally domesticated wheats devoted roughly 20 percent of their photosynthate to the development of seeds. Through plant breeding, it has been possible to raise this share—known as the "harvest index"—in today's wheat, rice, and corn to more than 50 percent. Given the essential requirements of the roots, stems, and leaves, the theoretical limit of the share going to seed is 60 percent.[7]

The key to this shift was the incorporation of the dwarf gene into both rice and wheat varieties by the Japanese during the late nineteenth century. Traditional wheat and rice varieties had tall straw because their wild ancestors needed to compete with other plants for sunlight. But once farmers began controlling weeds, growing tall was a waste of the plant's metabolic energy. As plant breeders shortened wheat and rice plants, reducing stem length, they reduced the share of photosynthate going into the straw and increased the portion going into seed. L. T. Evans, a prominent Australian agricultural scientist, observes that in the high-yielding dwarf wheats the gain in grain yield is roughly equal to the loss in straw weight from the dwarfing.[8]

With corn, varieties grown in the tropics were reduced in height from an average of nearly three meters to less than two. But Don Duvick, for many years the director of research at the Pioneer Hybrid seed company, observes that with the hybrids used in the U.S. Corn Belt, the key to higher yields is the ability of varieties to "withstand the stress of higher plant densities while still making the same amount of grain per plant." Growing more plants per hectare benefited from reorienting the horizontally inclined leaves of traditional strains that droop somewhat, making them more upright and thereby reducing the amount of self-shading.[9]

Although plant breeders have greatly increased the share of the photosynthate going to the seed, they have not been able to fundamentally improve the efficiency of photosynthesis—the process plants use to convert solar energy into biochemical energy. The amount of photosynthate produced from a given leaf area by today's crops remains unchanged from that of their wild ancestors.[10]

On the agronomic front, raising land productivity has depended on expanding irrigation, using more fertilizer, and controlling diseases, insects, and weeds. All these tactics help plants realize their genetic potential more fully.[11]

Other factors affecting yields include solar intensity and day length, natural conditions over which farmers have little control. Japan, for example, has developed a highly productive rice culture, one based on the precise spacing of rice plants in carefully tended rows. Yet rice yields in Spain, California, and Australia are consistently 20–30 percent higher. The reason is simple. These locations have an abundance of bright sunlight, whereas in Japan rice is necessarily grown during the monsoon season, when there is extensive cloud cover.[12]

Day length can also make a huge difference. To begin with, there are no high yields of any cereals—wheat, rice, or corn—in the equatorial regions. High yields come with the long growing days of summer in the higher latitudes. The world's highest wheat yields are found in Western Europe.[13]

Western Europe occupies a northerly latitude comparable to that of Canada and Russia, but the warmth from the Gulf Stream makes its winters mild, enabling the region to grow winter wheat. This wheat, planted in the fall, becomes well established and reaches several inches in height before winter dormancy begins. When early spring comes, it immediately begins to grow again. This enables wheat in Western Europe to mature during the summer, with days that are particularly long at that latitude. Thus four environmental conditions—moderate winters, inherently fertile soils, reliable rainfall, and long summer days—combine to give the region wheat yields that reach 6–8 tons per hectare.[14]

The difference in wheat yields among leading producers worldwide is explained more by soil moisture variations than by any other variable. Table 4–1 illustrates this point well. Kazakhstan, a country with low rainfall, aver-

Table 4–1. *Wheat Yield Per Hectare in Key Producing Countries, 2002*[1]

Country	Tons per Hectare
France	6.8
Mexico	5.0
China	3.8
India	2.7
United States	2.7
Canada	2.0
Argentina	2.2
Australia	1.7
Russia	1.8
Kazakhstan	1.1

[1]Yield for 2002 is the average of 2001 through 2003.
Source: See endnote 15.

ages 1.1 tons of wheat per hectare. France, the major wheat-growing country in Western Europe, harvests 6.8 tons per hectare—six times as much.[15]

Mexico's wheat yields are nearly double those of the United States primarily because virtually all of Mexico's wheat is irrigated, whereas the U.S. crop is largely rain-fed and grown in low rainfall regions. Similarly with India and Australia: in 1950, the yield in each country was roughly 1 ton of wheat per hectare. Today India gets 2.7 tons per hectare, while Australia gets only 1.7 tons. The reason is not that India's farmers are much more capable than farmers in Australia but rather that they can irrigate their wheat and can thus also efficiently use more fertilizer.[16]

Fertilizer and Irrigation

In 1847 Justus von Liebig, a German chemist, discovered that all the nutrients that plants remove from the soil could be replaced in chemical form. This insight had little immediate impact on agriculture, partly because growth in world food production during the nineteenth century came primarily from expanding cultivated area. It was not until the mid-twentieth century, when land limitations emerged, that fertilizer use began to climb.[17]

The rapid climb came as the frontiers of agricultural settlement disappeared and as the world began to urbanize quickly after World War II. With little new land to plow, growth in the food supply depended largely on raising crop yields. And this required more nutrients than were available in most soils. When the world was largely rural, plant nutrients were recycled as both human and livestock wastes were returned to the land. But with urbanization, this natural nutrient cycle was disrupted.

The shift from expanding cropland area to raising cropland productivity, coupled with accelerating urbanization, set the stage for the growth of the modern fertilizer industry. It also laid the groundwork for researchers to do elaborate soil testing to determine precisely which nutrients farmers needed to apply and when. It enabled farmers to remove nutrient constraints on yields, thus helping plants to realize their full genetic potential.

The growth in the world fertilizer industry after World War II was spectacular. Between 1950 and 1989, fertilizer use climbed from 14 million to 146 million tons. This period of remarkable worldwide growth came to an end when fertilizer use in the former Soviet Union fell precipitously after heavy subsidies were removed in 1988 and fertilizer prices there moved to world market levels. After 1990, the breakup of the Soviet Union and the effort of its former states to convert to market economies led to a severe economic depression in these transition economies. The combined effect of these shifts was a four-fifths drop in fertilizer use in the former Soviet Union between 1988 and 1995. After 1995 the decline bottomed out, and increases in other countries, particularly China and India, restored growth in world fertilizer use. (See Figure 4–1.)[18]

Among the big-three grain producers, China is the leading user of fertilizer, with the United States a distant second. India is now closing the gap with the United States and may overtake it within the next several years. (See Figure 4–2.)[19]

In many agriculturally advanced countries, fertilizer use has plateaued. For example, U.S. fertilizer use is essentially the same today as it was in the early 1980s, between 17 million and 21 million tons a year. Usage has also plateaued in Western Europe and Japan and will soon do the same in China as well.[20]

There are still some countries with a large potential for expanding fertilizer use. One is Brazil, which is not only raising land productivity but also steadily expanding

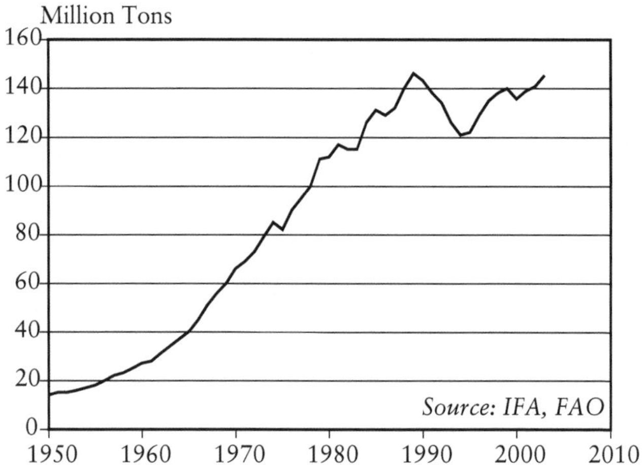

Figure 4–1. *World Fertilizer Use, 1950–2003*

the cultivated area. These two trends, together with the need to heavily fertilize nutrient-poor soils in both the cerrado and the Amazon basin, should continue the steady rise in fertilizer use in Brazil for the indefinite future. (The risks associated with this are discussed in Chapter 9.)[21]

For the world as a whole, however, the era of rapidly growing fertilizer use is now history. In the many countries that already have effectively removed nutrient constraints on crop yields, applying more fertilizer has little effect on yields. Indeed, where fertilizer application exceeds crop needs, nutrient runoff can contaminate drinking water and feed algal blooms that lead to eutrophication and offshore dead zones.[22]

Paralleling the tenfold increase in fertilizer use during the last half of the last century was the near tripling of irrigated area. (See Figure 4–3.) During the earlier part of this period, growth in irrigation came largely from the building of dams to store surface water and channel it

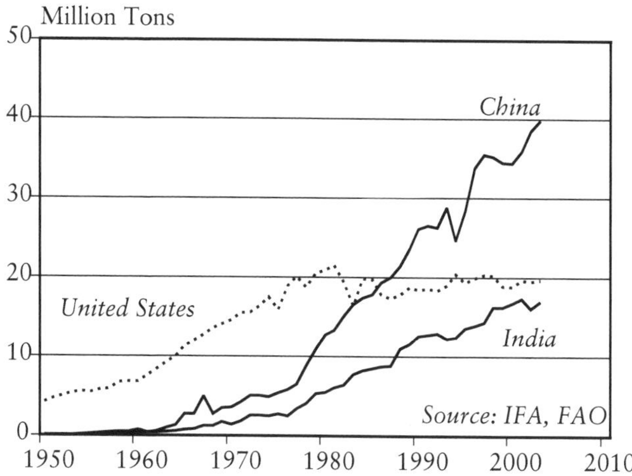

Figure 4–2. *Fertilizer Use by Country, 1950–2003*

onto the land through networks of gravity-fed surface canals. By the late 1960s, however, as the number of undeveloped dam sites diminished, farmers in countries like India and China were turning to underground water sources. Millions of irrigation wells were drilled during the remainder of the century.[23]

Now the potential for building new dams is limited. So, too, is that for drilling more irrigation wells, simply because the pumping volume of existing wells is already approaching or exceeding the sustainable yield of aquifers in key agricultural regions.

Over half of the world's irrigated land is in Asia, and most of that is in China and India. Some four fifths of China's grain harvest comes from irrigated land. This includes virtually all the riceland and most of the wheatland, plus part of the cornland. In India, over half of the grain harvest comes from irrigated land. And in the United States, irrigated land accounts for one fifth of the grain harvest.[24]

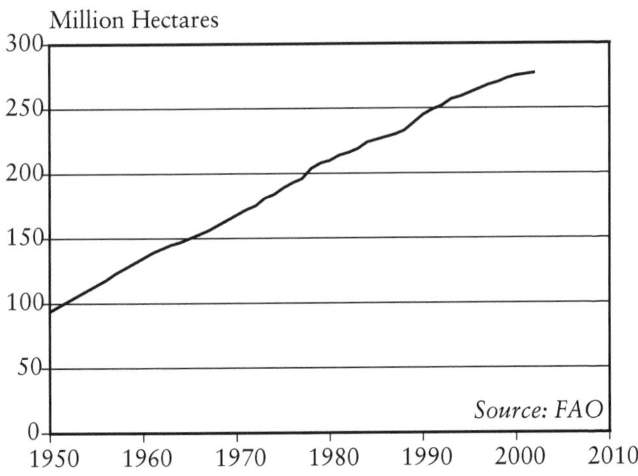

Figure 4–3. *World Irrigated Area, 1950–2002*

The growth in irrigation facilitated the growth in fertilizer use. Without irrigation in arid and semiarid regions, low soil moisture limits nutrient uptake and yields. When released of this constraint, plants can effectively use much more fertilizer. The availability of fertilizer makes investments in irrigation more profitable. It is this synergy between the growth in irrigation and fertilizer use that accounts for much of the world grain harvest growth over the last half-century or so.[25]

The availability of fertilizer helped to offset the loss of nutrients associated with the steadily expanding one-way flow of farm products, and the nutrients they contained, from farms to distant cities and other countries. The United States, for example, selling up to 100 million tons of grain a year to other countries, exports 2–3 million tons of the nutrients essential for plant growth, including nitrogen, phosphorus, and potassium. The use of chemical fertilizers prevents the outflow of grain from draining the croplands of the U.S. Corn Belt of nutrients.[26]

With irrigation as with fertilizer use, the growth worldwide has slowed dramatically over the last decade or so. Indeed, in some countries, such as Saudi Arabia and China, irrigated area is now shrinking. This is also true for parts of the United States, such as the southern Great Plains. In many parts of the world the need for water is simply outgrowing the sustainable supply.[27]

The Shrinking Backlog of Technology

Although the investment level in agricultural research, public and private, has not changed materially in recent years, the backlog of unused agricultural technology to raise land productivity is shrinking. In every farming community where yields have been rising rapidly, there comes a time when the rise slows and eventually levels off. For wheat growers in the United States and rice growers in Japan, for example, most of the available yield-raising technologies are already in use. Farmers in these countries are looking over the shoulders of agricultural researchers in their quest for new technologies to raise yields further. Unfortunately, they are not finding much.

From 1950 to 1990 the world's grain farmers raised the productivity of their land by an unprecedented 2.1 percent a year, slightly faster than the 1.9 annual growth of world population during the same period. But from 1990 to 2000 this dropped to 1.2 percent per year, scarcely half as fast. (See Table 4–2.) As of mid-2004, it looks as though the annual rise in grain yields from 2000 to 2010 will drop to something like 0.7 percent, scarcely half that of the preceding decade and far behind world population growth. This loss of momentum in raising land productivity is due not only to the shrinking backlog of technology but also in some countries to the loss of irrigation water.[28]

As noted earlier, yields vary widely among countries. This can be seen, for example, with the rice yields in

Table 4–2. *World Grain Yield Per Hectare, 1950–2000, with Projection to 2010*

Year	Yield Per Hectare[1]	Annual Increase By Decade
	(tons)	(percent)
1950	1.06	
1960	1.29	2.0
1970	1.65	2.5
1980	2.00	1.9
1990	2.47	2.1
2000	2.79	1.2
2010[2]	2.99	0.7

[1]Yields for decadal years 1960 through 2000 are three-year averages.
[2]Projection of yield to 2010 by author.
Source: See endnote 28.

Figure 4–4. Japan's rice yields, already quite high by 1960, appear to have plateaued over the last decade. The sharp drop in 1994 was the result of an unusually cool, cloudy monsoon season when solar intensity was well below normal.[29]

While India has doubled its rice yields over the last 40 years, they are still less than half those of China and Japan, and they have increased little over the last decade. This is partly because India's proximity to the equator means that it does not have the long summer days of Japan and China, both temperate-zone countries. And since scarcely half of India's rice production is irrigated, the remainder is entirely dependent on the vagaries of monsoon rainfall.[30]

The big story has been the advance in rice yields in China since the economic reforms in 1978. Now that China's rice yields are close to those of Japan's, however, which are the highest in Asia, it will become progressive-

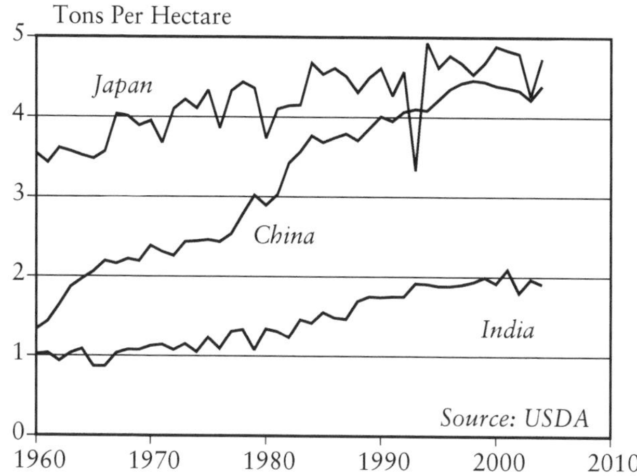

Figure 4–4. *Rice Yields in Japan, China, and India, 1960–2004*

ly more difficult to raise them further.[31]

Yields of wheat, the other principal food staple, also vary widely. U.S. wheat yields, though they fluctuate from year to year, have not increased much over the last two decades. (See Figure 4–5.) China's, in contrast, rose rapidly after the 1978 economic reforms but have shown signs of plateauing in recent years. In France, with some of the highest wheat yields in the world, yields also appear to have plateaued over the last decade or so.[32]

The corn yields in the world's three largest corn producers vary widely. (See Figure 4–6.) For instance, Brazil's are scarcely a third those of the United States. Much of China's corn is grown as a second crop after winter wheat, which means it gets planted up to several weeks later than the maximum yield planting time. By the time the corn has germinated, day length is already beginning to shorten.[33]

Figure 4–6 also shows how much corn production in the United States can fluctuate as a result of heat and

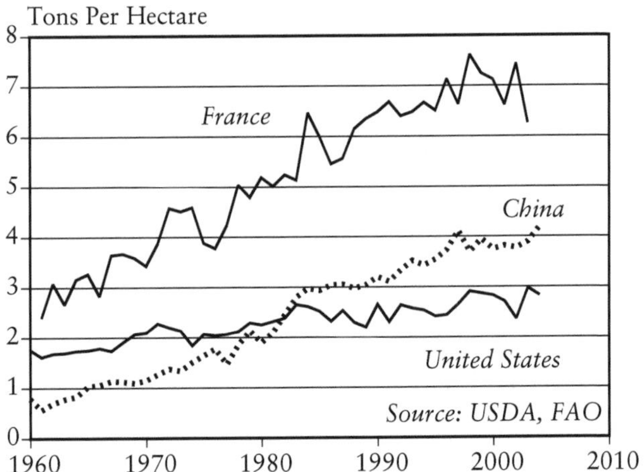

Figure 4–5. *Wheat Yields in France, China, and the United States, 1960–2004*

drought. The two big drops, 1983 and 1988, were both associated with intense heat and drought. In both 2003 and 2004, exceptionally favorable weather helped boost yields well above the trend.[34]

In addition to plateauing in agriculturally advanced countries such as Japan and South Korea, rice yields are also stagnating in several developing countries in Asia. In an analysis of yield trends and potentials, Kenneth Cassman and his colleagues at the University of Nebraska point out that in three of China's major rice-producing provinces, which account for 35 percent of the country's harvest, yields are stagnating.[35]

In India, the world's second largest rice producer after China, yields are leveling off in the Punjab, where wheat and rice are extensively double-cropped. Signs of rice yield plateaus are also appearing in Indonesia's Central Java and in central Luzon, the largest island in the Philippines.[36]

The Nebraska team further notes that there was no

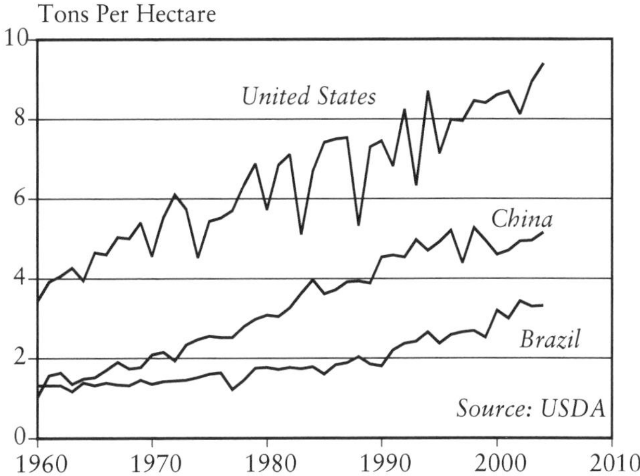

Figure 4–6. *Corn Yields in the United States, China, and Brazil, 1960–2004*

detectable yield gain in inbred rice varieties during the 37 years since the development of IR-8, an early prototype of the high-yielding rices from the 1960s, at IRRI. The only gains since then have come from hybrid rices, which China has led the way on. But these hybrids yield only 9 percent more than the much more widely grown inbred varieties. Half of China's rice area is now planted to hybrids, but the area has not increased for many years, partly because hybrid rices are plagued by high seed cost and poor grain quality.[37]

In 1990 IRRI launched a major research project to raise rice yields 25–50 percent by restructuring the rice plant. In the face of poor prospects for achieving this, the goal has now been scaled back to a rise of 5–10 percent.[38]

In looking at the potential for raising wheat yields in developing countries, Cassman and colleagues note that wheat yields also appear to be stagnating in Mexico's Yaqui Valley, the site of the international wheat-breeding effort that over the last 60 years produced the widely adapted versions of the high-yielding Japanese dwarf wheats that were at the heart of the Green Revolution.[39]

In the Indian states of Punjab and Haryana, the country's leading producers of irrigated wheat, yields are approaching those where the leveling off began in the Yaqui Valley. Since these two states account for 34 percent of India's wheat harvest, reaching a plateau in yields here would substantially slow the rise in the national harvest trend.[40]

For maize, the Nebraska team looked at the results of an irrigated maize yield competition for Nebraskan corn growers and noted the winning yield had not increased for 20 years. In other words, no varietal improvement or agronomic advances have enabled the contest winners to raise their yields. The Nebraska statewide average corn yield on all farms is continuing to rise on both irrigated and non-irrigated land, as is the yield of the contest winners on non-irrigated land.[41]

Can genetic engineers restore a rapid worldwide rise in grainland productivity? This prospect is not promising simply because plant breeders using traditional techniques have largely exploited the genetic potential for increasing the share of photosynthate that goes into seed. Once this is pushed close to its limit, the remaining options tend to be relatively small, clustering around efforts to raise the plant's tolerance of various stresses, such as drought or soil salinity. One major option left to scientists is to increase the efficiency of the process of photosynthesis itself—something that has thus far remained beyond their reach.

After 20 years of research, biotechnologists have yet to produce a single variety of wheat, rice, or corn that would dramatically raise yields above those of existing varieties. Thus far the focus in genetically engineered crops has been to develop herbicide tolerance, insect resistance, and disease resistance. Between 1987 and 2001, 70 percent of the applications for field releases of experimental genetically engineered varieties received by the USDA's Animal and Plant Health Inspection Service, the regulatory agency for genetically modified crops, were in these three areas. Some 27 percent of the requested releases were for herbicide-tolerant varieties, principally soybeans. The second highest category, insect resistance, accounted for 25 percent of the total, including cotton varieties resistant to the boll weevil and corn varieties resistant to the corn borer. Crop varieties resistant to various diseases caused by viruses, fungi, or bacteria together accounted for 18 percent of new releases.[42]

Some 6 percent of the requested releases had specific agronomic properties, such as drought resistance or salt tolerance, while 17 percent were focused on improving crop quality in some particular way. The latter category included crop strains that contained a specific trait such as higher protein quality in corn or higher oil content in soybeans. Not one of these varieties was bred to raise yields. To the extent that insect- and disease-resistant varieties provide better pest control than the use of pesticides, this could marginally increase crop output. But as a general matter, yield gains thus far from biotechnology are minimal to non-existent.[43]

When genetic yield potential is close to the physiological limit, further advances in yields rely on exploiting the remaining unrealized potential in the use of basic inputs, such as fertilizer and irrigation, or on the fine-tuning of other agronomic practices, such as optimum planting densities or more effective pest controls. Beyond this, there will eventually come a point in each country, with each grain, when farmers will not be able to sustain the rise in yields.

USDA plant scientist Thomas R. Sinclair observes that advances in our understanding of plant physiology let scientists quantify crop yield potentials quite precisely. He notes that "except for a few options which allow small increases in the yield ceiling, the physiological limit to crop yields may well have been reached under experimental conditions." For farmers who are using the highest-yielding varieties that plant breeders can provide, along with the agronomic inputs and practices needed to realize their genetic potential, there may be few options left to raise land productivity.[44]

Reinforcing this view is the work cited earlier by Kenneth Cassman and colleagues that notes stagnation in raising the genetic yield potential of the major cereal crops—rice and maize, when average yields reach 80 percent of the genetic yield potential. Cassman points out that it is difficult to raise them further because "achieving 100 percent of the genetic yield potential requires perfect management in terms of varietal selection, plant density, planting date, nutrient management (neither deficiency or excess and perfect balance amongst all 16 essential nutrients), and in the control of weeds, insects, and diseases." He notes that average farm yields tend to plateau at 80–85 percent of the genetic yield potential.[45]

Most countries that have achieved a yield takeoff have managed at least to double if not triple or even quadruple grain yields. Among those that have quadrupled yields over the past half-century are the United States and China with corn; France, the United Kingdom, and Mexico with wheat; and China with rice. The bottom line is that all countries are drawing on a backlog of shrinking agricultural technology. And for some crops in some countries the backlog has largely disappeared.[46]

The decelerating rise in grain yields since 1990 is not peculiar to individual grains or individual countries. It reflects a systemic difficulty in sustaining the gains that characterized the preceding four decades as yields of wheat, rice, and corn press against the ceiling ultimately imposed by the limits of photosynthetic efficiency. The efficiency of photosynthesis coupled with the area of land available to produce food defines the outer limit of how much food the earth can produce.

Future Options

In a world where it is becoming increasingly difficult to raise land productivity, we have to look for alternative ways of expanding output. One obvious approach is to increase the amount of multiple cropping—growing more than one crop on a field per year. Yet this is not easy, and in some East Asian countries, such as Japan, South Korea, Taiwan, and, more recently, China, it is already declining.[47]

Devising economic incentives to sustain multiple cropping in some countries and expand it in others could help buy time to stabilize world population size. For countries in East Asia, the challenge is to provide economic incentives to farmers so as to avoid, or at least slow, the decline in double cropping. In the United States, in contrast, where the overriding concern for half a century was to control production by restricting the area planted to grain, the potential for more multiple cropping may be surprisingly large. Here economic incentives for double cropping could boost output. One of the keys to exploiting this lies in reorienting agricultural research programs to develop facilitating technologies such as earlier maturing crops and farm practices that will accelerate the harvesting of the first crop and the planting of the second one.

Another way to expand food production is to raise water productivity. This helps both to preserve the existing irrigated area, where water supplies are tightening, and to expand the area irrigated in other places. The water available for irrigation can also be increased at the local level by building small water-harvesting ponds. These not only capture rainfall runoff, holding it for irrigation, they also help recharge underground aquifers.

Land productivity can be raised by using crop residues to produce food. For example, the tonnage of wheat straw, rice straw, and corn stalks produced worldwide easily matches the weight of the grain produced by these crops. As India has demonstrated with its world leadership in milk production, and as China is showing with its surging beef production, it is now possible to feed these vast quantities of crop residues to animals, converting them into milk and meat. In effect, this permits a "second harvest" from the same land.[48]

In some parts of the world, such as Africa, investment in transportation and storage infrastructure can play a major role in boosting food production, enabling farmers to move beyond subsistence agriculture. This is particularly helpful in both getting inputs such as fertilizer to farmers and getting their harvest to markets.[49]

Jules Pretty, director of the Centre for Environment and Society at the University of Essex, has pioneered a broad concept of sustainable agriculture, one that strives to develop natural, human, and social capital at the local level. It emphasizes the use of local resources. Sustainable farming, says Pretty, "seeks to make the best use of nature's goods and services. It minimizes the use of nonrenewable inputs (pesticides and fertilizers) that damage the environment.... It makes better use of the knowledge and skills of farmers."[50]

In reviewing the results of some 45 sustainable agriculture initiatives in 17 African countries, Pretty notes that both crop yields and nutritional levels improved more or less apace. Overall, he notes that crop yields are up 50–100 percent in these projects over 20 years.[51]

Included in the sustainable agriculture toolbox is the better use of local natural resources and processes like nutrient cycling, nitrogen fixation, soil rebuilding, and the use of natural enemies to control pests. This approach does not rule out the use of fertilizer and pesticides but seeks to minimize the need for their use. The use of leguminous plants to supply nitrogen is seen as an intrinsic part of the process. Animal manures are collected to fertilize fields and build up soil organic matter. This, in turn, increases soil moisture retention.

The emphasis on human capital leads to greater self-reliance by farmers. Learning centers and extension offices play an important role in the communities with successful sustainable agriculture. With social capital, the key is getting people to work together, in groups, to better manage watersheds and local forests or to supply credit to small-scale farmers.

With this approach, communities with marginal land have succeeded not only in raising incomes and improving diets, but also in producing a marketable surplus of farm products. Highly successful though this approach is, it does require substantial support to energize local communities. Pretty notes that "without appropriate policy support, [these community projects] are likely to remain localized in extent, and at worst simply wither away."[52]

The challenge is to raise land productivity in one way or another and to design research programs to do this while protecting the land and water resource base and avoiding damage to natural systems, such as that caused by nutrient runoff.

5

Protecting Cropland

On April 18, 2001, the western United States—from the Arizona border north to Canada—was blanketed with dust. The dirt came from a huge dust storm that originated in northwestern China and Mongolia on April 5. Measuring 1,800 kilometers across when it left China, the storm carried up to 100 million tons of topsoil, a vital resource that would take centuries to replace through natural processes.[1]

Almost exactly one year later, on April 12, 2002, South Korea was engulfed by a huge dust storm from China that left people in Seoul literally gasping for breath. Schools were closed, airline flights were cancelled, and clinics were overrun with patients having difficulty breathing. Retail sales fell. Koreans have come to dread the arrival of what they now call "the fifth season," the dust storms of late winter and early spring.[2]

These two dust storms, among some 20 or more major dust storms in China during 2001 and 2002, are one of the externally visible indicators of the ecological catastrophe unfolding in northern and western China. Overgrazing and overplowing are converting productive land to desert on an unprecedented scale. Other dust storms are occurring in Africa, mostly in the southern

Sahara and the Sahelian zone. Scientists estimate that Chad alone may be exporting 1.3 billion tons of topsoil each year to the Atlantic Ocean, the Caribbean islands, and even Florida in the United States. Wind erosion of soil and the resulting desert creation and expansion are shrinking the cropland base in scores of countries.[3]

Another powerful pressure on cropland is the automobile. Worldwide, close to 400,000 hectares (1 million acres) of land, much of it cropland, are paved each year for roads, highways, and parking lots. In densely populated, low-income developing countries, the car is competing with farmers for scarce arable land.[4]

The addition of more than 70 million people each year requires land for living and working—driving the continuous construction of houses, apartment buildings, factories, and office buildings. Worldwide, for every 1 million people added, an estimated 40,000 hectares of land are needed for basic living space.[5]

These threats to the world's cropland, whether advancing deserts, expanding automobile fleets, or housing developments, are gaining momentum, challenging some of the basic premises on which current population, transportation, and land use policies rest.

Losing Soil and Fertility

Soil erosion is not new. It is as old as the earth itself. But with the advent of agriculture, the acceleration of soil erosion on mismanaged land increased to the point where soil loss often exceeded new soil formation. Once this threshold is crossed, the inherent fertility of the land begins to fall.

As soil accumulated over millennia, it provided a medium in which plants could grow. Plants protected the soil from erosion. The biological fertility of the earth is due to the accumulation of topsoil over long stretches of geologic time—the product of a mutually beneficial relationship between plants and soil. But as the human enterprise expanded, soil erosion began to exceed new soil formation in more and more areas, slowly thinning the layer of topsoil that had built up over time. Each year the world's farmers are challenged to feed another 70 million or more people but with less topsoil than the year before.[6]

Erosion of soil by water and wind reduces the fertility of rangeland and cropland. For the rangelands that support the nearly 3.1 billion head of cattle, sheep, and goats in our custody, the threat comes from the overgrazing that destroys vegetation, leaving the land vulnerable to erosion. Rangelands, located mostly in semiarid regions of the world, are particularly vulnerable to wind erosion.[7]

In farming, erosion comes from plowing land that is steeply sloping or too dry to support adequate soil protection with ground cover. Steeply sloping land that is not protected by terraces, by perennial crops, or some other way loses soil when it rains heavily. Thus the land hunger that drives farmers up mountainsides fuels erosion. Land that is excessively dry, usually receiving below 25 centimeters (10 inches) of rain a year, is highly vulnerable to wind erosion once vegetation, typically grass, is cleared for cropping or by overgrazing. Under cultivation, this soil often begins to blow away.[8]

In the United States, wind erosion is common in the semiarid Great Plains, where the country's wheat production is concentrated. In the U.S. Corn Belt, where most of the country's corn and soybeans are grown, the principal erosion threat is from water. This is particularly true in the states with rolling land and plentiful rainfall, such as Iowa and Missouri.[9]

Land degradation from both water and wind erosion in the world's vulnerable drylands is extensive, affecting some 900 million hectares (see Table 5–1), an area substantially larger than the world's grainlands (some 670

Table 5–1. *Soil Degradation by Region in Susceptible Drylands, 1990s*

	Water Erosion	Wind Erosion	Total
		(million hectares)	
North America	38	38	76
South America	35	27	62
Europe	48	39	87
Africa	119	160	279
Asia	158	153	311
Australasia	70	16	86
Total	468	433	901

Source: See endnote 10.

million hectares). Two thirds of this damaged land is in Africa and Asia, including the Middle East. These also are the world's two most populous regions. And they are where fully two thirds of the 3 billion people expected to be added to world population by 2050 will live. If more people translate into more livestock, as historically has been the case, the damage will spread to still more land.[10]

The enormous twentieth-century expansion in world food production pushed agriculture onto highly vulnerable land in many countries. The overplowing of the U.S. Great Plains during the late nineteenth and early twentieth century, for example, led to the 1930s Dust Bowl. This was a tragic era in U.S. history—one that forced hundreds of thousands of farm families to leave the Great Plains. Many migrated to California in search of a new life, a movement immortalized in John Steinbeck's *The Grapes of Wrath*.[11]

Three decades later, history repeated itself in the Soviet Union. The Virgin Lands Project, a huge effort to convert grassland into grainland between 1954 and 1960, led to the plowing of an area for wheat that exceeded the wheatland in Canada and Australia combined. Initially this resulted in an impressive expansion in Soviet grain production, but the success was short-lived as a dust bowl developed there too.[12]

Kazakhstan, at the center of the Virgin Lands Project, saw its grainland area peak and begin to decline around 1980. After reaching a historical high of just over 25 million hectares, it shrank to barely half that size—13 million hectares. Even on the remaining land, however, the average wheat yield is only 1.1 tons per hectare, a far cry from the nearly 7 tons per hectare that farmers get in France, Western Europe's leading wheat producer and exporter. This precipitous drop in Kazakhstan's grain harvest illustrates the price that other countries will have to pay for overplowing and overgrazing.[13]

In the closing decades of the twentieth century, yet another dust bowl—perhaps the biggest of all—began developing in China. As described in Chapter 8, it is the result of overgrazing, overplowing, overcutting of trees, and overpumping of aquifers, all of which make the land in northern and western China more vulnerable to erosion.[14]

Africa, too, is suffering from heavy losses of topsoil as a result of wind erosion. Andrew Goudie, Professor of Geography at Oxford University, reports that dust storms originating over the Sahara—once so rare—are now commonplace. He estimates they have increased tenfold during the last half-century. Among the countries most affected by topsoil loss via dust storms are Niger, Chad, northern Nigeria, and Burkino Faso. In Mauritania, in Africa's far west, the number of dust storms jumped from 2 a year in the early 1960s to 80 a year today.[15]

The Bodélé Depression in Chad is the source of an estimated 1.3 billion tons of dust a year, up tenfold from 1947, when measurements began. Dust storms leaving

Africa travel westward across the Atlantic, depositing so much dust in the Caribbean that they damage coral reefs there. When the dust is carried northward and deposited on Greenland, it reduces the reflectivity of the ice, leading to greater heat absorption and accelerated ice melting. The 2–3 billion tons of fine soil particles that leave Africa each year in dust storms are slowly draining the continent of its fertility and, hence, its biological productivity.[16]

Dust storms and sand storms are a regular feature of life in the Middle East as well. The Sistan Basin on the border of Afghanistan and Iran is now a common source of dust storms in that region. Once a fertile complex of lakes and marshes fed by the Helmand River, which originates in the highlands of eastern Afghanistan, the area has become largely a desert as the river has been drained dry by the increasing water withdrawals by Afghan farmers for irrigation.[17]

The bottom line is that the accelerating loss of topsoil from wind and water erosion is slowly but surely reducing the earth's inherent biological productivity. Unless governments, farmers, and herders can mobilize to reverse this trend, feeding 70 million more people each year will become progressively more difficult.

Advancing Deserts

Roughly one tenth of the earth's land surface is used to produce crops. Two tenths is grassland of varying degrees of productivity. Another two tenths is forest. The remaining half of the land is either desert, mountains, or covered with ice. The area in desert is expanding, largely at the expense of grassland and cropland. Deserts are advancing in Africa both north and south of the Sahara and throughout the Middle East, the Central Asian republics, and western and northern China. (See Table 5–2.) (The effect of desertification on China's food pro-

Table 5–2. *Selected Examples of Desertification Around the World*

Country	Extent of Desertification
Afghanistan	In the Sistan basin, windblown dust and sand have buried more than 100 villages. In the northwest, along the Amu Darya River, a sand dune belt that is some 300 kilometers (186 miles) long and 30 kilometers wide is expanding by up to 1 meter a day.
Brazil	Approximately 58 million hectares of land have been affected. Economic losses associated with desertification are estimated at $300 million a year.
China	Nationwide, deserts are expanding by 360,000 hectares a year. Some 400 million Chinese are affected by the dust storms of late winter and early spring.
India	Various forms of desertification affect 107 million hectares, one third of India's land area.
Iran	In the eastern provinces of Baluchistan and Sistan, some 124 villages have been buried by drifting sand.
Kenya	More than 80 percent of its land is vulnerable to desertification, affecting up to a third of the country's 32 million people and half its livestock.
Mexico	Some 70 percent of all land in Mexico is vulnerable to desertification. Land degradation prompts some 700,000 Mexicans to leave the land each year in search of jobs in Mexican cities or the United States.

Table 5–2. *continued*

Nigeria	Each year 351,000 hectares of land are lost to desertification, which affects each of the 10 northern states.
Yemen	Some 97 percent of the land in this country of 19 million people shows some degree of desertification.

Source: See endnote 18.

duction is discussed in more detail in Chapter 8.)[18]

Nigeria, Africa's most populous country, is losing 351,000 hectares of rangeland and cropland to desertification each year. While Nigeria's human population has increased from 30 million in 1950 to 130 million in 2004, a fourfold expansion, its livestock population has grown from roughly 6 million to 65 million head, a tenfold increase. With the forage needs of Nigeria's 15 million head of cattle and nearly 50 million sheep and goats exceeding the sustainable yield of the country's grasslands, the country is slowly turning to desert.[19]

The government of Nigeria considers the loss of productive land to desert to be far and away its leading environmental problem. No other environmental change threatens to undermine its economic future so directly as the conversion of productive land to desert. Conditions will only get worse if Nigeria continues on its current population trajectory toward 258 million people by 2050.[20]

In the vast swath of Africa between the Sahara Desert and the forested regions to the south lies the Sahel. In countries from Senegal and Mauritania in the west to Sudan, Ethiopia, and Somalia in the east, human and livestock pressures are converting more and more land into desert.[21]

A similar situation exists along the Sahara's northern edge, the tier of largely semiarid countries across the top

After sand storms, electricity and telephone poles are often buried in the drifting sand dunes. In order to keep systems working, soldiers have to raise the lines by extending the pole. Photo: Lu Tongjing.

of Africa. Algeria, in particular, is being squeezed between the Mediterranean Sea and the Sahara Desert as the latter advances northward. In a desperate effort to halt this encroachment, Algeria has decided to convert the southernmost 20 percent of its grainland to perennial crops, such as olive orchards or grape vineyards, that will hold the soil better. Whether this will halt the advancing desert remains to be seen. At a minimum, it will be a difficult sacrifice in a country that already imports 40 percent of its grain.[22]

After more sand storms, the accumulating sand is carried to another location, leaving the extended pole far above the lowered land surface. Photo: Lu Tongjing.

Some of the most severe desertification found anywhere is in China, where 360,000 hectares of land become desert each year. In parts of northern and western China, deserts have expanded to the point where they are beginning to merge. In China's Xinjiang Province, the huge Taklamakan and the smaller Kumtag deserts are approaching each other and appear headed for a merger. On the southwestern edge of Inner Mongolia, the 5-million-hectare (12-million-acre) Bardanjilin desert is moving toward the 3-million-hectare Tengry desert.[23]

In these regions of desert expansion, sandstorms are

common, often forcing the abandonment of villages. Keeping highways passable becomes a major challenge as sand dunes advance across roadways. Keeping power lines and telephone lines above the drifting sand is itself a challenge. Special crews periodically follow the phone lines across the countryside looking for poles that may be about to be inundated with drifting sand. They then extend the poles to make sure the lines remain above the sand. But a few months later, the sand may be blown away, leaving the wires suspended far above the ground.[24]

Converting Cropland to Other Uses

In addition to losing cropland to severe soil erosion and desert expansion, the world is also losing cropland to various nonfarm uses, including residential construction, industrial construction, the paving of roads and parking lots, and airports, as well as to recreational uses, such as tennis courts and golf courses. If for every million people added to the world's population, 40,000 hectares of land are needed for nonfarm uses, adding more than 70 million people each year claims nearly 3 million hectares, part of which is agricultural land. The cropland share of land converted to nonfarm uses varies widely both within and among countries, but since cities are typically located on the most fertile land, it is often high—sometimes 100 percent.[25]

China is currently working to create 100 million jobs in the manufacturing sector. With the average factory in China employing 100 workers, China needs to build 1 million factories—many of which will be sited on former cropland. India, with the annual addition of 18 million people and with accelerating economic growth, is facing similar pressures to convert cropland to other uses.[26]

Residential building claims on cropland are also heavy. If we assume each dwelling houses on average five people, then adding 70 million or more people to world population each year means building 14 million houses or apartments annually.[27]

While population growth spurs housing demand, rising incomes spur automobile ownership. The world automobile fleet is expanding by roughly 9 million per year. (See Figure 5–1.) Each car requires the paving of some land, with the amount paved ranging from a high of 0.07 hectares per vehicle in sparsely populated countries such as the United States, Canada, or Brazil to a low of 0.02 hectares in densely populated areas such as Europe, Japan, China, and India.[28]

As long as a fleet is growing, the country has no choice but to pave more land if it wants to avoid gridlock. In India, a country of only 8 million cars, each new million cars require the paving of roughly 20,000 hectares of land. If it is cropland, and of average productivity, this translates into roughly 50,000 tons of grain, enough to feed 250,000 people at the country's current meager food con-

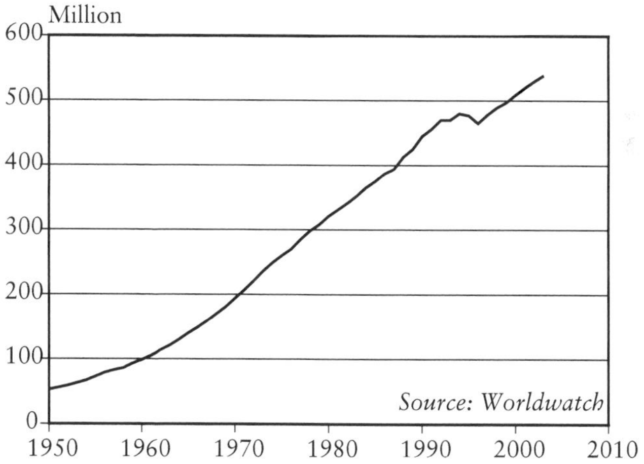

Figure 5–1. *World Automobile Fleet, 1950–2003*

sumption level. A country that will need to feed an additional 515 million people by 2050 cannot afford to cover scarce cropland with asphalt for roads and parking lots.[29]

As the world's affluent turn to the automobile, they are competing for land with those who are hungry and malnourished. Governments in developing countries are essentially using their financial resources to underwrite the public infrastructure for the automobile often at the expense of the hungry.

In the United States, where 0.07 hectares of paved land is required for each car, every five cars added to the fleet require paving an area the size of a football field. Thus the 2 million cars added to the U.S. fleet each year require asphalting an area equal to nearly 400,000 football fields.[30]

Just parking the 214 million motor vehicles owned by Americans requires a vast area of land. Imagine a parking lot with a fleet of 214 million vehicles. If that is difficult, try visualizing a parking lot for 1,000 cars, and then imagine 214,000 such parking lots. The 16 million hectares (61,000 square miles) of U.S. land devoted to roads, highways, and parking lots compares with 21 million hectares that American farmers planted in wheat in 2004.[31]

As the new century gets under way, the competition between cars and crops for land is heating up. Until recently the paving over of cropland has occurred largely in industrial countries, where four fifths of the world's 539 million automobiles are found. But now more and more farmland is being sacrificed in developing countries with hungry populations, calling into question the future role of the car.[32]

There is not enough land in China, India, and other densely populated countries like Indonesia, Bangladesh, Pakistan, Iran, Egypt, and Mexico to both support automobile-centered transportation systems and feed people. The competition between cars and crops for land is thus becoming a struggle between the rich and the poor—between those who can afford automobiles and those who are struggling to get enough food to survive.

Conserving Topsoil

In contrast to the loss of cropland to nonfarm uses, which is often beyond the control of farmers, the losses of soil and eroded land from severe erosion are within their control. Reducing soil losses caused by wind and water erosion to below the rate of new soil formation will take an enormous worldwide effort. Based on the experience of leading food producers such as China and the United States, as well as numerous smaller countries, easily 5 percent of the world's cropland is highly erodible and should be converted back to grass or trees before it becomes wasteland. The first step to halting the decline in inherent land fertility is to pull back from this fast-deteriorating margin.[33]

The key to controlling wind erosion is to keep the land covered with vegetation as much as possible and to slow wind speeds at ground level. Ground-level wind speeds can be slowed by planting shrubs or trees on field borders and by leaving crop residues on the surface of the soil. For areas with strong winds and in need of electricity, such as northwestern China, wind turbines can simultaneously slow wind speeds and provide cheap electricity. This approach converts an agricultural liability—strong winds—into an economic asset.

One time-tested method for dealing with water erosion is terracing, as is so common in rice paddies throughout the mountainous regions of Asia. On less steeply sloping land, contour strip farming, as found in the U.S. Midwest, works well.[34]

Another tool in the soil conservation toolkit—and a relatively new one—is conservation tillage, which includes both no-till and minimum tillage. Farmers are learning that less tillage may be better for their crops. Instead of the traditional cultural practices of plowing land, discing or harrowing it to prepare the seedbed, and then planting with a seeder and cultivating row crops with a mechanical cultivator two or three times to control weeds, farmers simply drill seeds directly through crop residues into undisturbed soil. Weeds are controlled with herbicides. The only soil disturbance is the creation of a narrow slit where the seeds are inserted just below the surface.[35]

This practice, now widely used in the production of corn and soybeans in the United States, has spread rapidly in the western hemisphere over the last two decades. (See Table 5–3.) Data for crop year 2003/04 show the United States with nearly 24 million hectares of land under no-till. Brazil had nearly 22 million hectares, Argentina 16 million, and Canada 13 million. Australia's 9 million hectares rounds out the five leading no-till countries.[36]

In the United States, the combination of retiring highly erodible land under the landmark Conservation Reserve Program that began in 1985, and required farmers to develop conservation plans on cropland eroding excessively, has sharply reduced soil erosion. In addition to the no-till cropland, 19 million more hectares were minimum-tilled, for a total of 43 million hectares of conservation tillage. Conservation tillage was used on 37 percent of the corn crop, 57 percent of the soybean crop, and 30 percent of wheat and other small grains.[37]

Once farmers master the practice of no-till, its use can spread rapidly, particularly if governments provide economic incentives or require farm soil conservation plans for farmers to be eligible for crop subsidies. In the United States, the no-till area went from 7 million hectares in 1990 to nearly 24 million hectares in 2003/04, more than tripling. Recent FAO reports describe the growth in no-

Table 5–3. *Cropland Under No-Till in Key Countries, 2003/04*

Country	Area (million hectares)
United States	23.7
Brazil	21.9
Argentina	16.0
Canada	13.4
Australia	9.0
Paraguay	1.5
Pakistan/Northern India	1.5
Bolivia	0.4
South Africa	0.3
Spain	0.3
Venezuela	0.3
Uruguay	0.3
France	0.2
Chile	0.1
Others	1.2
Total	90.1

Source: See endnote 36.

till farming over the last few years in Europe, Africa, and Asia. In addition to reducing both wind and water erosion, and particularly the latter, this practice also helps retain water, raises soil carbon content, and reduces the energy needed for crop cultivation.[38]

Saving Cropland

Every person added to the world's population in a land-scarce time provides another reason for protecting cropland from conversion to nonfarm uses. Ideally, we would build our homes, offices, factories, shopping malls, roads, and parking lots only on land that is unsuitable for farming. Unfortunately, people are concentrated where the best cropland is located—either because they are farmers or because land that is good for crops is typically the flat, well-drained land that is also ideal for cities and the construction of roads.

This reality underscores the importance of land use planning in the development of human settlements and also in the formulation of transportation policy. The U. S. sprawl model of development is not only land-intensive, it is also energy-inefficient and aesthetically unappealing. Urban sprawl leaves people trapped in communities not densely populated enough to support a first-class public transport system, thus forcing them to commute by car, with all the attendant congestion, pollution, and frustration.

Automobiles promised mobility, and in largely rural societies they provided it. But in urban situations a continually increasing number of cars eventually brings immobility. There is an inherent conflict between the car and the city. After a point, the more cars, the less mobility. Some cities are now taxing cars every time they enter a city or the center district. Initially Singapore but now also Melbourne, Oslo, and, most recently, London have adopted this disincentive to encourage commuters to shift to public transportation, which takes much less land.[39]

European governments, which have followed a very different development model from the United States, have carefully zoned their urban development, leading to a much more land-efficient, energy-efficient, aesthetically pleasing approach. Ironically, Americans often spend their vacations biking in the English or French countrysides, so they can enjoy picturesque rural settings not destroyed by sprawl.

In developing countries facing acute land scarcity, there is now another pressing reason for protecting cropland from the automobile and urban sprawl. China, for example, has been blindly following the western industrial development model. In 1994, it announced that it was going to develop an auto-centered transportation system, inviting manufacturers such as Toyota, General Motors, and Volkswagen to submit proposals for building assembly plants in China.[40]

Within a matter of months a group of senior Chinese scientists, including members of the Academy of Sciences, had produced a white paper challenging this decision. They noted the oil import needs this policy entailed, along with the traffic congestion and air pollution. But their principal question was whether China had enough land both to feed its people and to support an auto-centered transportation system. Their conclusion was that it did not and that the government should build an alternative urban transportation model that used far less land—a system centered on light rail, buses, and bicycles.[41]

We are indebted to these scientists for recognizing early on that the automobile-centered, western industrial development model is simply not appropriate for densely populated developing countries. Nor over the long term is this model likely to be viable in industrial countries either. Numerous European cities are not only investing in first-rate public transportation systems, they are also actively encouraging the use of bicycles for travel within the city. Amsterdam and Copenhagen, where up to 40 percent of all trips within the city are taken by bicycle, are leading the way.[42]

When industrial countries were rapidly urbanizing during the twentieth century, agricultural land was considered a surplus commodity. Now it is a scarce resource.

In today's densely populated developing countries, the amount of land used by transportation systems directly affects food production. In a world of 6 billion people, transportation policy and food security are intimately related.[43]

Data for figures and additional information can be found at www.earth-policy.org/Books/Out/index.htm.

6

Stabilizing Water Tables

Although public attention has recently focused on the depletion of oil resources, the depletion of underground water resources poses a far greater threat to our future. While there are substitutes for oil, there are none for water. Indeed, we lived for millions of years without oil, but we would live for only a matter of days without water.

Not only are there no substitutes for water, but we need vast amounts of it to produce food. At the personal level, we drink roughly four liters of water a day (nearly four quarts), either directly or indirectly in various beverages. But it takes 2,000 liters of water—500 times as much—to produce the food we consume each day.[1]

Since food is such an extraordinarily water-intensive product, it comes as no surprise that 70 percent of world water use is for irrigation. Although it is now widely accepted that the world is facing water shortages, most people have not yet connected the dots to see that a future of water shortages will also be a future of food shortages.[2]

Falling Water Tables

Over much of the earth, the demand for water exceeds the sustainable yield of aquifers and rivers. The gap between the continuously growing use of water and the sustainable supply is widening each year, making it more and more difficult to support rapid growth in food production.[3]

With river water in key farming regions rather fully exploited, the world has turned to underground water sources in recent decades to keep expanding the irrigated area. As a result, the climbing demand for water has now exceeded the natural recharge of many aquifers. Now water tables are falling in scores of countries that contain more than half the world's people. (See Table 6–1.) These include China, India, and the United States, which together account for nearly half of the global grain harvest. And as the gap between steadily rising demand and the sustainable yield of aquifers grows, water tables are falling at an accelerating rate.[4]

In the United States, water tables are falling under the Great Plains and throughout the southwest. In India, they are falling in most states—including the Punjab, the nation's breadbasket. This country of more than 1 billion people depends on underground water supplies for well over half of its irrigation water, with the remainder coming from rivers. In China, water tables are falling throughout the northern half of the country, including under the North China Plain, the source of half of the nation's wheat and a third of its corn.[5]

The effects of aquifer depletion vary, depending on whether it is a replenishable or fossil aquifer. If the aquifer is replenishable, as most are, once depletion occurs the water pumped is necessarily reduced to the amount of recharge. If, for example, an aquifer is being pumped at twice the rate of recharge, depletion means the rate of pumping will be cut in half. In a fossil, or nonreplenishable aquifer, however, depletion means the end of pumping. Fossil aquifers include the Ogallala under the U.S. Great Plains, the aquifer the Saudis use to irrigate wheat, and the deeper of the two aquifers under the North China Plain.[6]

Table 6–1. *Underground Water Depletion in Key Countries*

Country	Description
Mexico	In Mexico, where a third of all water comes from underground, aquifers are being depleted throughout the northern arid and semiarid regions. In a country where irrigated land is more than three times as productive as rain-fed land, the loss of irrigation water from aquifer depletion will be costly.
United States	Overpumping is widespread, and the overpumping of the vast Ogallala or High Plains aquifer—essentially a fossil aquifer that extends from southern South Dakota through Nebraska, Kansas, eastern Colorado, Oklahoma, and Texas—is a matter of national concern. In the southern Great Plains, irrigated area has shrunk by 24 percent since 1980 as wells have gone dry.
Saudi Arabia	When the Saudis turned to their large fossil aquifer for irrigation, wheat production climbed from 140,000 tons in 1980 to 4.1 million tons in 1992. But with rapid depletion of the aquifer, production dropped to 1.6 million tons in 2004. It is only a matter of time until irrigated wheat production ends.
Iran	The overpumping of aquifers is estimated at 5 billion tons per year. When aquifers are depleted, Iran's grain harvest could drop by 5 million tons, or one third of the current harvest.
Yemen	This country of 21 million people is unique in that it has both one of the world's fastest-growing populations and the fastest-falling

Table 6–1. *continued*

	water tables. The World Bank reports that the water table is falling by 2 meters or more a year in most of Yemen.
Israel	Both the coastal aquifer and the mountain aquifer Israel shares with Palestinians are being depleted. With severe water shortages leading to a ban on irrigated wheat, the continuous tightening of water supplies is likely to further raise tensions in the region.
India	Water tables are falling in most states in India, including the Punjab and Haryana, the leading grain-surplus states. With thousands of irrigation wells going dry each year, India's farmers are finding it increasingly difficult to feed the 18 million people added each year.
China	Water tables are falling throughout northern China, including under the North China Plain. China's harvest of wheat has fallen in recent years as irrigation wells have dried up. From 2002 to 2004, China went from being essentially self-sufficient in wheat to being the world's largest importer.

Source: See endnote 4.

In some countries, falling water tables and the depletion of aquifers are already reducing the grain harvest. In Saudi Arabia, the wheat harvest peaked in 1992 at 4.1 million tons, and then declined to 1.6 million tons in 2004—a drop of 61 percent. (See Figure 6–1.) Some other smaller countries, such as Yemen, have also experienced grain harvest declines.[7]

For the first time, diminishing supplies of irrigation water are helping to shrink the grain harvest in a large grain producer—China. The wheat harvest, which peaked at 123 million tons in 1997, dropped to 90 million tons in 2004—a decline of 27 percent. The production of wheat has dropped much more than that of corn and rice because wheat is grown largely in the semiarid northern half of the country, where water is scarce.[8]

Serious though emerging water shortages are in China, the problem may be even more serious in India simply because the margin between actual food consumption and survival is so precarious. In a recent survey of India's water situation, Fred Pearce in the *New Scientist* notes that the 21 million wells drilled in this global epicenter of well-drilling are lowering water tables in most of the country. The wells, powered by heavily subsidized electricity, are dropping water tables at an accelerating rate. In North Gujarat, the water table is falling by

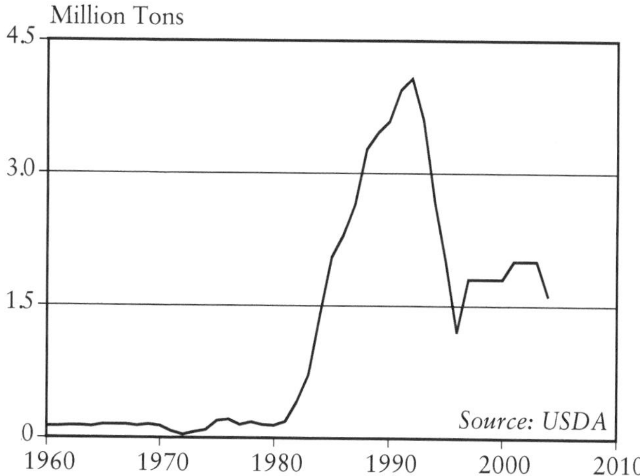

Figure 6–1. *Wheat Production in Saudi Arabia, 1960–2004*

6 meters or 20 feet per year. In some states, half of all electricity is now used to pump water.[9]

In Tamil Nadu, a state of 62 million people in southern India, falling water tables are drying up wells. Kuppannan Palanisami of Tamil Nadu Agricultural University says that falling water tables have dried up 95 percent of the wells owned by small farmers, reducing the irrigated area in the state by half over the last decade.[10]

As water tables fall, well drillers are using modified oil-drilling technology to reach water, drilling as deep as 1,000 meters in some locations. In communities where underground water sources have dried up entirely, all agriculture is rain-fed and drinking water is trucked in. Tushaar Shah, who heads the International Water Management Institute's groundwater station in Gujarat, says of India, "When the balloon bursts, untold anarchy will be the lot of rural India."[11]

In Mexico, water tables are falling throughout the more arid north. As this happens, the energy required to pump water rises. Indeed, more than 6 percent of Mexico's electricity is used to pump water.[12]

In the United States, the loss of irrigation water is making it more difficult for farmers to respond to the future import needs of other countries. In the southern Great Plains, for example, the irrigated area has shrunk by 24 percent since 1980. Leading agricultural states such as Texas, Oklahoma, and Kansas are among those most affected by falling water tables.[13]

In a rational world, falling water tables would trigger alarm, setting in motion a series of government actions to reduce demand and reestablish a stable balance with the sustainable supply. Unfortunately, not a single government appears to have done this. Official responses to falling water tables have been consistently belated and grossly inadequate.

Rivers Running Dry

While falling water tables are largely invisible, rivers that are drained dry before they reach the sea are highly visible. Two rivers where this phenomenon can be seen are the Colorado, the major river in the southwestern United States, and the Yellow, the largest river in northern China. Other large rivers that either run dry or are reduced to a mere trickle during the dry season are the Nile, the lifeline of Egypt; the Indus, which supplies most of Pakistan's irrigation water; and the Ganges in India's densely populated Gangetic basin. (See Table 6–2.)[14]

Some rivers have disappeared entirely. A few years ago, China announced plans to divert water from the Yellow River to Taiyuan, the capital of Shaanxi Province. Learning of this, I asked why they did not simply take water from the Fen, the local river that originated in northern Shaanxi and flowed southward through Taiyuan, eventually emptying into the Yellow River on the province's southern border. Fred Crook, senior China analyst at the U.S. Department of Agriculture, responded that the Fen River had dried up. It exists only on old maps. Taiyuan is now wholly dependent on underground water resources from well fields, and some of the wells are starting to go dry.[15]

The management and use of scarce water in river basins that include several countries can be difficult. The Nile, for instance, which originates largely in Ethiopia and flows through Sudan and Egypt, is reduced to a trickle by the time it reaches the Mediterranean Sea. Since it rarely rains in Egypt, the country's existence depends entirely on the Nile.[16]

Virtually all the river's water is now being used. Against this backdrop, governments could be expected to coordinate population policy with water availability, but there appears to be no effort to do so. In Egypt, the population is projected to grow from 73 million today to 127 million in 2050. Sudan's population is projected to nearly double, from 34 million today to 60 million. Ethiopia, meanwhile, is projected to go from 72 million today to 171 million by 2050.[17]

Egypt now gets the lion's share of the Nile's water partly because it developed much sooner than Ethiopia. But as Ethiopia begins to develop, it is planning to build dams on the upper (Blue) Nile that will reduce the flow in the lower reaches of the Nile river basin. It will be difficult for Egypt, where incomes average $3,900 per year, to argue that Ethiopia, with incomes of only $710 per year, should

Table 6–2. *Major Rivers Running Dry*

River	Condition
Amu Darya	The Amu Darya, which originates in the mountains of Afghanistan, is one of the two rivers that feed into the Aral Sea. Soaring demands on this river, largely to support irrigated agriculture in Uzbekistan, sometimes drain it dry before it reaches the sea. This, along with a reduced flow of the Syr Darya—the other river feeding into the sea—helps explain why the Aral Sea has shrunk by more than half over the last 40 years.
Colorado	All the water in the Colorado, the major river in the southwestern United States, is allocated. As a result, this river, fed by the rainfall and snowmelt from the mountains of Colorado, now rarely makes it to the Gulf of California.
Fen	This river, which flowed from the northern part of China's Shaanxi province and emptied into the Yellow River at the province's southern end, has literally disappeared as water withdrawals upstream in the watershed have dropped the water table, drying up springs that once fed the river.

not be permitted to develop its water resources. With virtually all the water in the basin now spoken for and with the combined population of the three countries projected to grow from 179 million to 358 million by 2050, the potential for the basin's population to outgrow its water resources—setting the stage for conflict—is clear.[18]

Another major river where conflicts over water rights are arising is the Mekong. China's construction of several huge hydroelectric dams on the upper reaches of the river system is reducing the Mekong's flow, directly affecting fisheries, navigation, and irrigation prospects downstream in Cambodia, Laos, and Viet Nam.[19]

Table 6–2. *continued*

Ganges	Some 300 million people of India live in the Ganges basin. Flowing through Bangladesh en route to the Bay of Bengal, the Ganges has little water left when it reaches the bay.
Indus	The Indus, originating in the Himalayas and flowing west to the Indian Ocean, feeds Pakistan's irrigated agriculture. It now barely reaches the ocean during much of the year. Pakistan, with a population of 157 million projected to reach 349 million by 2050, is facing trouble.
Nile	In Egypt, a country where it rarely ever rains, the Nile is vitally important. Already reduced to a trickle when it reaches the Mediterranean, it may go dry further upstream in the decades ahead if, as projected, the populations of Sudan and Ethiopia double by 2050.
Yellow	The cradle of Chinese civilization, the Yellow River frequently runs dry before it reaches the sea.

Source: See endnote 14.

Yet potential flashpoint is the Amu Darya, a river that originates in Afghanistan and flows through Turkmenistan and Uzbekistan before reaching the Aral Sea. As upstream Afghanistan stabilizes politically and begins to develop, it plans to claim some of the water from the river, which will reduce the amount available to the two downstream countries.[20]

We do not know whether sharing water in the Tigris-Euphrates River basin, where irrigated agriculture began some 6,000 years ago, was a source of conflict historically in the region. But today it is a source of tension between Turkey, Syria, and Iraq. In recent years, Turkey has invested heavily in a network of dams, mostly in the upper reaches of the Tigris, that are providing power and water for a large irrigation expansion. Turkey is one of the few countries in the world with a major expansion of irrigation still under way. Again, population growth is contributing to the mounting tensions in this river basin, since the populations of both Syria and Iraq are projected to double by mid-century.[21]

There are literally dozens of other shared rivers where the demand for water is now pressing against the limits of supply, forcing countries to work out agreements on the allocation of river water. Once these agreements are reached, it is in each country's interest to use its share of the water as efficiently as possible.

Cities Versus Farms

At the international level, water conflicts among countries dominate the headlines. But within countries it is the competition for water between cities and farms that preoccupies political leaders. Neither economics nor politics favors farms. They almost always lose out to cities.

In many countries farmers are now faced with not only a shrinking water supply but also a shrinking share of that shrinking supply. In large areas of the United States, such as the southern Great Plains and the Southwest, virtually all water is now spoken for. Meanwhile, the demand for water continues to climb in the region's fast-growing cities, including Denver, Phoenix, Las Vegas, Los Angeles, and San Diego. The growing water needs of these cities and of thousands of small towns in the region can be satisfied only by taking water from agriculture.[22]

A California monthly magazine, *The Water Strategist*, devotes several pages in each issue to a listing of water sales in the western United States during the preceding month. Scarcely a day goes by without the announcement of a new sale. Eight out of ten are by individual farmers or their irrigation districts to cities and municipalities.[23]

Colorado, with a fast-growing population, has one of the world's most active water markets. Cities and towns of all size are buying irrigation water rights from farmers and ranchers. In the Arkansas River basin, which occupies the southeastern quarter of the state, Colorado Springs and Aurora (a suburb of Denver) have already bought water rights to one third of the basin's farmland. Aurora has purchased rights to water that was once used to irrigate 9,600 hectares (23,000 acres) of cropland in the Arkansas valley.[24]

Far larger purchases are being made by cities in California, a state of 36 million people. In 2003, San Diego bought annual rights to 247 million cubic meters (1 cubic meter of water equals 1 ton) of water from farmers in nearby Imperial Valley—the largest rural/urban water transfer in U.S. history. This agreement covers the next 75 years. In 2004, the Metropolitan Water District, which supplies water to 18 million southern Californians in several cities, negotiated the purchase of 137 million cubic meters of water per year from farmers for the next 35 years. Without irrigation water, the highly productive land owned by these farmers is wasteland. The farmers who are selling their water rights would like to continue to grow crops, but city officials are offering much more for the water than the farmers could possibly earn by using it to produce crops.[25]

In many countries, farmers are not compensated for a loss of irrigation water. In 2004, for example, Chinese farmers along the Juma River downstream from Beijing discovered that the river had run dry. A diversion dam had been built near the capital to take river water for Yanshan Petrochemical, a state-owned industry. Although there were bitter protests by the farmers, it was a losing battle. For the 120,000 villagers downstream from the diversion dam, livelihoods would suffer, perhaps crippling their ability to make a living from farming. Whether it is a result of outright government expropriation, farmers being outbid by cities, or cities simply drilling deeper wells than farmers can afford, the world's farmers are losing the water war.[26]

In the competition between cities and farms, cities have the advantage simply because they can pay much more for water. In China, a thousand tons of water can be used to produce 1 ton of wheat, worth at most $200, or it can be used to expand industrial output by $14,000—70 times as much. In a country where industrial development and the jobs associated with it are an overriding national economic goal, scarce water is no longer going to farmers. Agriculture is becoming the residual claimant on the world's increasingly scarce supply of water.[27]

Scarcity Crossing National Boundaries

Historically, water shortages were local, but today scarcity is crossing national boundaries via the international grain trade. As just described, countries facing shortages divert water from irrigation to satisfy the growing demand in cities, and then import grain to offset the loss of farm output. The reason for this is simple: since it takes a thousand tons of water to produce a ton of grain, the most efficient way to import water is as grain. In effect, countries are using grain as a currency to balance their water books. Trading in grain futures is in a sense trading in water futures.[28]

The fastest-growing grain import market in the world in recent years has been North Africa and the Middle East. The demand for grain there is growing quickly as a result of both rapid population growth and rising affluence, much of it derived from the export of oil. Virtually every country in this region is pushing against the limits of local water supplies. To meet growing water needs in cities, governments routinely divert irrigation water from agriculture.[29]

Last year the water required to produce the grain and other farm products brought into the region equaled the annual flow of the Nile River. In effect, the water deficit in the region can be thought of as another Nile flowing into North Africa and the Middle East in the form of imported grain.[30]

It has been fashionable in recent years to say that future wars in the Middle East are more likely to be fought over water than over oil. But it is not only costly to win a water war, it is difficult to secure water supplies by winning. In reality, the wars over water are taking place in world grain markets. It is the countries that are financially strongest—not those that are militarily strongest—that will prevail in this competition.

Raising Water Productivity

To avoid a water crunch that leads to a food crunch requires a worldwide effort to raise water productivity. The tightening water situation today is similar to what the world faced with land a half-century ago. After World War II, as governments assessed the food prospect for the remainder of the century, they saw both enormous projected growth in world population and little new land to bring under the plow. In response, they joined with international development institutions in a worldwide effort to raise land productivity that was replete with commodity price supports, heavy investment in agricultural research, extension services, and farm credit agencies. The result was a rise in world grainland productivity from 1.1 tons per hectare in 1950 to 2.9 tons in 2004.[31]

Today the world needs to launch a similar effort to raise water productivity. Land productivity is measured in tons of grain per hectare or bushels per acre, but there are no universally used indicators to measure and discuss water productivity. The indicator likely to emerge for irrigation water is kilograms of grain produced per ton of water. Worldwide that average is now roughly 1 kilogram of grain per ton of water used.[32]

The first challenge is to raise the efficiency of irrigation water, since this accounts for 70 percent of world water use. Some data have been compiled on water irrigation efficiency at the international level for surface water projects—that is, dams that deliver water to farmers through a network of canals. Water policy analysts Sandra Postel and Amy Vickers write about a 2000 review that found that "surface water irrigation efficiency ranges between 25 and 40 percent in India, Mexico, Pakistan, the Philippines, and Thailand; between 40 and 45 percent in Malaysia and Morocco; and between 50 and 60 percent in Israel, Japan, and Taiwan." Irrigation water efficiency is affected not only by the mode and condition of irrigation systems but also by soil type, temperature, and humidity. In arid regions with high temperatures, the evaporation of irrigation water is far higher than in humid regions with lower temperatures.[33]

In a May 2004 meeting, China's Minister of Water

Resources Wang Shucheng outlined for me in some detail plans to raise China's irrigation efficiency from 43 percent in 2000 to 51 percent in 2010 and then to 55 percent in 2030. The steps he described to boost irrigation water efficiency included raising the price of water, providing incentives for adopting more irrigation-efficient technologies, and developing the local institutions to manage this process. Reaching these goals, he said, would assure China's future food security.[34]

Crop usage of irrigation water never reaches 100 percent simply because some irrigation water evaporates from the land surface, some percolates downward, and some runs off. When attempting to raise the water efficiency of irrigation, the trend is to shift from the less efficient flood-or-furrow system to overhead sprinkler irrigation or to drip irrigation, the gold standard of irrigation water efficiency. Low-pressure sprinkler systems reduce water use by an estimated 30 percent over flood or furrow irrigation, while switching from flood or furrow to drip irrigation typically cuts water use in half.[35]

As an alternative to furrow irrigation, a drip system also raises yields because it provides a steady supply of water with minimal losses to evaporation. Since drip systems are both labor-intensive and water-efficient, they are well suited to countries with underemployment and water shortages. They allow farmers to raise their water productivity by using labor, which is often in surplus in rural communities.[36]

Recent data indicate that a few small countries—Cyprus, Israel, and Jordan—rely heavily on drip irrigation to water their crops. (See Table 6–3.) Among the big three agricultural producers—China, India, and the United States—the share of irrigated land using these more-efficient technologies ranges from less than 1 percent in India and China to 4 percent in the United States.[37]

Table 6–3. *Use of Drip and Micro-irrigation, Selected Countries, Circa 2000*

Country	Area Irrigated by Drip and Other Micro-irrigation Methods[1]	Share of Total Irrigated Area Under Drip and Micro-irrigation
	(thousand hectares)	(percent)
Cyprus	36	90
Israel	125	66
Jordan	38	55
South Africa	220	17
Spain	563	17
Brazil	176	6
United States	850	4
Chile	62	3
Egypt	104	3
Mexico	143	2
China	267	<1
India	260	<1

[1]Micro-irrigation typically includes drip (both surface and subsurface) methods and micro-sprinklers; year of reporting varies by country.
Source: See endnote 37.

Low water productivity is often the result of low water prices. Current water prices are often irrationally low, belonging to an era when water was an abundant resource. As water becomes scarce, it needs to be priced accordingly. In Beijing, public hearings were held in mid-2004 on a proposal to raise water prices. At the end of July, officials announced rate hikes for urban and industrial users of some 26 percent, effective August 1. The price went from 4.01 yuan (48¢) to 5.04 yuan (61¢) per cubic meter. Other local governments in northern China, mostly at the provincial level, have been raising water prices in small increments to discourage waste. The advantage of higher prices is that it affects the decisions of all water users. Higher prices encourage investment in more water-efficient irrigation technologies, industrial processes, and household appliances.[38]

In many cities in water-short parts of the world, it may be time to rethink the typical urban water use model, one where water flows into the city, is used once, and then leaves the city—usually becoming polluted in the process. This flush-and-forget model that so dominates urban water systems will not be viable over the longer term in water-scarce regions. One alternative sewage system is the use of so-called dry toilets, which do not use water and which convert human waste into a rich humus, a highly valued fertilizer.

Another variation on the existing urban water use models is one that comprehensively recycles urban water supplies. Water can be used indefinitely in cities and by industry if it is recycled. Some cities are beginning to do this. Singapore, for example, which buys its water from Malaysia, is starting to recycle its water in order to reduce this vulnerable dependence.[39]

Some countries can realize large water savings by restructuring the energy sector, shifting from fossil-fuel-powered thermal plants, which require large amounts of water for cooling, to renewable energy sources, such as wind and solar. In the United States, for instance, the 48 percent of total water withdrawals that is used for thermal cooling exceeds the 34 percent withdrawn for irrigation. Most of the water used for thermal cooling is river water that returns to its source once it is used, albeit much warmer than when it was withdrawn. Although the actual water losses from evaporation in the power plant cooling towers typically amount to only 7 percent of the water that goes through the plants, the return of the hot water to the river is often ecologically damaging.[40]

What is needed now is a new mindset, a new way of thinking about water use. In addition to more-efficient irrigation technologies, for example, shifting to more water-efficient crops wherever possible also boosts water productivity. Rice production is being phased out in the region around Beijing because it is so water-intensive. Similarly, Egypt restricts rice production in favor of wheat.[41]

Anything that raises the productivity of irrigated land typically raises the productivity of irrigation water. Anything that increases the efficiency with which grain is converted into animal protein increases water productivity.

For people consuming excessive amounts of livestock products, moving down the food chain means not only a healthier diet and reduced health care costs, but also a reduction in water use. In the United States, where the consumption of grain as food and feed averages some 800 kilograms (four fifths of a ton) per person, a modest reduction in eating livestock products could easily cut grain use per person by 100 kilograms. Given that there are 297 million Americans, such a reduction would cut grain use by 30 million tons and the use of water to produce grain by 30 billion tons. At average world grain consumption levels of roughly 300 kilograms per person a year, 30 million tons of grain would feed 100 million people—more than enough to cover world population growth for one year.[42]

Reducing water use to a level that can be sustained by aquifers and rivers worldwide involves a wide range of measures not only in agriculture but also throughout the economy. Among some of the more obvious steps are shifting to more water-efficient irrigation practices and technologies, planting more water-efficient crops, adopting more water-efficient industrial processes, and using more water-efficient household appliances. One of the less conventional steps is to shift from outdated coal-fired power plants, which require vast amounts of water for thermal cooling, to wind power—something long overdue in any case for reasons of pollution and climate disruption. Recycling urban water supplies is another obvious step to consider in countries facing acute water shortages.

The need to stabilize water tables is urgent, thanks to the sheer geographic scale of overpumping, the simultaneity of falling water tables among countries, and the accelerating drop in water level. Although falling water tables are historically a recent phenomenon, they now threaten the security of water supplies and, hence, of food supplies in countries containing 3.2 billion people. Beyond this, the shortfall—the gap between the use of water and the sustainable yield of aquifers—grows larger each year, which means the water level drop is greater than the year before. Underlying the urgency of dealing with the fast-tightening water situation is the sobering realization that not a single country has succeeded in stopping the fall in its water tables and stabilizing water levels. The fast-unfolding water crunch has not yet translated into food shortages, but if unaddressed, it may soon do so.[43]

Data for figures and additional information can be found at www.earth-policy.org/Books/Out/index.htm.

7

Stabilizing Climate

In July of 2004, the U.S. National Academy of Sciences released a research report by a team of nine scientists from China, India, the Philippines, and the United States who had measured the precise effect of rising temperatures on rice yields under field conditions. They concluded that yields typically fall by 10 percent for each 1-degree Celsius rise in temperature during the growing season. This confirmed what had seemed obvious to many agricultural analysts, namely that high temperatures can shrink harvests.[1]

In recent years, numerous heat waves have lowered grain harvests in key food-producing countries. In 2002, record-high temperatures and associated drought reduced grain harvests in India, the United States, and Canada, dropping the world harvest 89 million tons below consumption. In 2003, Europe was hit by high temperatures. The record-breaking late summer heat wave that claimed 35,000 lives in eight nations shrank harvests in every country from France eastward through the Ukraine. It contributed to a world harvest shortfall of 94 million tons—5 percent of world consumption.[2]

The new research results from agricultural scientists, along with the grain production performance of various countries recently exposed to record temperatures, underscore the close relationship between energy policy and food security. Farmers already struggling to feed 70 million or more people each year will find it even more difficult as the earth's temperature rises.[3]

Rising Temperatures, Falling Yields

Within just the last few years, crop ecologists in several countries have been focusing on the precise relationship between temperature and crop yields. In an age of rising temperatures, their findings are disturbing. One of the most comprehensive of these studies was the one just cited, which focused on rice yields. This study was conducted at the International Rice Research Institute (IRRI) in the Philippines, the world's premier rice research organization. The IRRI team of eminent crop scientists noted that from 1979 to 2003, the annual mean temperature at the research site rose by roughly 0.75 degrees Celsius.[4]

Using crop yield data from the experimental field plots for irrigated rice under optimal management practices for the years 1992–2003, the team's finding confirmed the rule of thumb emerging among crop ecologists—that a 1-degree-Celsius rise in temperature lowers wheat, rice, and corn yields by 10 percent. The IRRI finding was consistent with those of other recent research projects. They concluded that "temperature increases due to global warming will make it increasingly difficult to feed Earth's growing population."[5]

While this study analyzing rice yields was under way, an empirical historical analysis of the effect of temperature on corn and soybean yields was being conducted in the United States. It concluded that higher temperatures had an even greater effect on yields of these crops. Using data for 1982–98 from 618 counties for corn and 444 counties for soybeans, David Lobell and Gregory Asner concluded that for each 1-degree Celsius rise in temperature, yields declined by 17 percent. Given the projected temperature increases in the U.S. Corn Belt, where a large share of the world's corn and soybeans are produced, these findings should be of grave concern to those responsible for world food security.[6]

The most vulnerable part of the crop cycle is the pollination period that immediately precedes seed formation. One of the IRRI projects, for example, showed that at 34 degrees Celsius (93 degrees Fahrenheit), nearly 100 percent of the tiny flowers on a rice head turn into kernels of rice. But at 40 degrees Celsius (104 degrees Fahrenheit), only a few kernels develop, leading to crop failure.[7]

Wheat and corn are similarly vulnerable. Earlier research showed that higher carbon dioxide (CO_2) levels in the atmosphere led to higher grain yields, assuming that there are no constraints imposed by soil moisture, nutrient availability, or other limiting factors. What the new research shows is that the negative effect of higher temperature on crop yields overrides the positive effect of higher CO_2 levels. Indeed, if pollination fails and there is no seed formation, then the CO_2 effect on grain yield is lost entirely.[8]

Abnormally high temperatures directly affect yields by stressing crops. Anyone who has been in a cornfield in mid-summer with temperatures above 35 degrees Celsius has seen how tightly the leaves curl in order to reduce moisture loss. But this also reduces photosynthesis, often to the point where the corn plant is merely maintaining itself. Under conditions of intense heat, plant growth ceases entirely.[9]

As temperatures rise, crop-withering heat waves are becoming more and more common. On August 12, 2003, when the U.S. Department of Agriculture released its monthly estimate of the world grain harvest, it reported a 32-million-ton drop from the July estimate. This drop, equal to half the U.S. wheat harvest, was concentrated in Europe, where record-high temperatures had withered crops in virtually every country in the region.[10]

The heat wave in Europe began in early summer 2003, when Switzerland experienced the hottest June since recordkeeping began 140 years ago. In July, the heat engulfed nearly the whole continent. In late summer, soaring temperatures were rewriting the European record book. On August 10th, the temperature in London reached 38 degrees Celsius (100 degrees Fahrenheit). France had 11 consecutive days in August with temperatures above 35 degrees Celsius. In Italy, temperatures reached 41 degrees Celsius.[11]

Crops suffered the most in Eastern Europe, which harvested its smallest wheat crop in 30 years. The wheat crop in the Ukraine, already severely damaged by winterkill, was reduced further by the heat, plummeting from 21 million tons the year before to a mere 5 million tons. As a result, the Ukraine—a leading wheat exporter in 2002—was forced to import wheat in late 2003 and early 2004 as bread prices threatened to spiral out of control. Romania, which was particularly hard hit by heat and drought, harvested the smallest wheat crop on record. And the Czech Republic had its poorest grain harvest in 25 years.[12]

During this life-threatening heat wave, Europeans may have felt that the temperature could not rise much more. But the increases projected for the decades ahead mean that such events will become more frequent and more intense. Just as Europeans could not have imagined the severity of the heat wave in the summer of 2003 that claimed 35,000 lives and shrank grain harvests in virtually every country, so too we have difficulty visualizing the extreme heat waves yet to come.[13]

Temperature Trends and Effects

Since 1970, the earth's average temperature has risen by 0.7 degrees Celsius, or nearly 1.3 degrees Fahrenheit. Each decade the rise in temperature has been greater than in the preceding one. (See Figure 7–1.) Four of the six warmest years since recordkeeping began in 1880 have come in the last six years. Two of these, 2002 and 2003, were years in which, as just described, the major food-producing regions saw their crops wither in the presence of record or near-record temperatures.[14]

As atmospheric concentrations of CO_2 rise, so does the earth's temperature. Since atmospheric CO_2 permits sunlight to freely penetrate the earth's atmosphere but restricts the radiation of heat back into space, it creates a "greenhouse effect."

Atmospheric concentrations of CO_2, estimated at 280 parts per million when the Industrial Revolution began, have been rising ever since people in Europe began burning coal. (See Figure 7–2.) They have risen every year

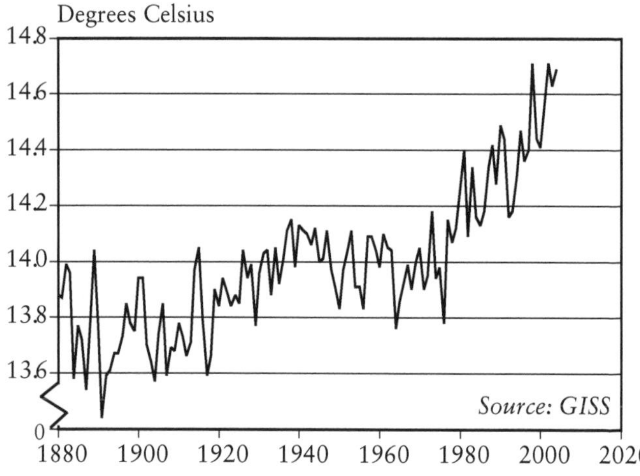

Figure 7–1. *Average Global Temperature, 1880–2003*

since precise measurements began in 1959, making this one of the world's most predictable environmental trends. As shown in Figure 7–2, atmospheric CO_2 concentrations turned sharply upward around 1960. Roughly a decade later, around 1970, the temperature too began to climb; the rise since then is quite visible in Figure 7–1. Projections by the Intergovernmental Panel on Climate Change (IPCC) show temperatures rising during this century by 1.4–5.8 degrees Celsius. The accelerating rise in temperature in recent years appears to have the world headed toward the upper end of that projected range of increase.[15]

Perhaps even more important than the average temperature rise is where the increase is likely to be concentrated. The warming will be greater over land than over the oceans, in the higher latitudes than in the equatorial regions, and in the interior of continents than in the coastal regions. One of the higher increases is expected to be in the interior of North America—an area that

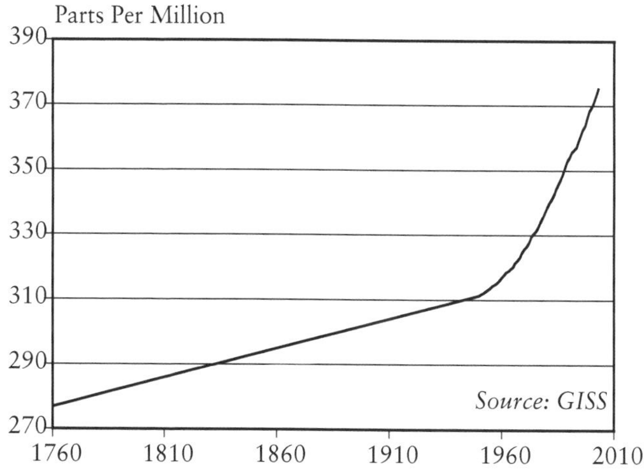

Figure 7–2. *Global Atmospheric Concentrations of Carbon Dioxide, 1760–2003*

includes the grain-growing Great Plains of the United States and Canada and the U.S. Corn Belt, the very region that makes this continent the world's breadbasket.[16]

The earth's rising temperature affects food security in many ways. Much of the world's fresh water is stored in ice and snow in mountainous regions. These "reservoirs in the sky" supply water for irrigation. But the reservoirs are now shrinking. A modest rise in temperature of 1 degree Celsius in mountainous regions can substantially alter the precipitation mix between rain and snow, increasing rainfall and decreasing snowfall. This leads to more runoff during the rainy season and less snowmelt to feed rivers during the dry season, when farmers need irrigation water.[17]

The melting glaciers and shrinking snowfields of the Himalayas are a concern to countries throughout Asia because this is where virtually all the major rivers in the region originate—the Indus, Ganges, Mekong, Yangtze, and Yellow. In Asia, where half the world's people live and where irrigated agriculture looms large, any reduction in river flow during the summer directly affects food security. The prospect of diminished river flows during the dry season at a time when water tables are already falling in most Asian countries raises basic questions about food security in the region.[18]

In addition to the direct effects of temperature on yield, higher temperatures mean more evaporation and thus more rainfall. Elevated temperatures can lead both to more extreme drought and to more severe flooding. Drought can be caused by below-normal rainfall or above-normal temperatures. Most often the two combine to create crop-withering droughts. Increased temperatures also mean more powerful, more destructive storms.[19]

Higher temperatures can worsen or create new crop disease and insect problems. The combination of heat and humidity, which makes an ideal environment for many plant diseases, makes it almost impossible to produce wheat profitably in the tropics. Higher temperatures would simply expand the region that is inhospitable to wheat from the equator toward the higher latitudes.[20]

One of the most serious long-term effects of climate change is rising sea level, which is driven both by the thermal expansion of the oceans as temperatures rise and by the melting of glaciers. The last IPCC report projected that sea level could rise by up to one meter during the current century, but papers published since then indicate that the melting is proceeding much faster than IPCC scientists had estimated. One study of glaciers in Alaska and Western Canada, for example, suggests that ice melting there is now raising sea level by 0.32 millimeters per year, more than double the 0.14 millimeters per year assumed by IPCC.[21]

One of the major concerns among scientists today is the accelerated melting of the Greenland ice sheet. If the ice sheet on Greenland—an island three times the size of Texas—were to melt entirely, sea level would rise 7 meters (23 feet), inundating not only Asia's rice-growing river deltas and floodplains but most of the world's coastal cities as well. This kind of massive melting, even in the case of the most rapid warming scenario, would occur over centuries, however, not years.[22]

The World Bank has published a map of Bangladesh, which shows that a 1-meter rise in sea level would inundate half of the country's riceland. It would also displace some 40 million Bangladeshis. Where would these people go? Which countries would be willing to accept even a million refugees fleeing the effects of rising sea level?[23]

A warmer earth means that agricultural zones in the northern hemisphere would move northward within Canada and Russia, for example, as the growing season lengthens. This assumes, of course, that there are high-quality soils that could sustain a productive agriculture in these regions. In Canada, however, the glaciated soils north of the Great Lakes cannot begin to match the productivity of the deep, fertile U.S. Corn Belt soils south of the Great Lakes.[24]

One advantage of a longer growing season would be that the winter wheat belt could move northward, replacing the lower-yielding spring wheat now grown in the northernmost agricultural regions. This would affect primarily Canada and Russia, the leading producers of spring wheat.[25]

On balance, however, agriculture would be a heavy loser if temperature continues to rise. The notion that the world's farmers would be better off with more atmospheric CO_2 and higher temperatures is a view based more on wishful thinking than on science. It may soon become apparent that the costs of climate change are unacceptably high.

Raising Energy Efficiency

If rising temperatures continue to shrink harvests and begin driving up food prices, public pressure to stabilize climate by cutting the carbon emissions that cause the greenhouse effect could become intense. The goal is to cut these emissions enough to stabilize climate and eliminate the threat to world food security from rising temperatures. Cutting emissions enough to stabilize atmospheric CO_2 levels is an ambitious undertaking, but given the technologies now available to both raise energy efficiency and develop renewable sources of energy, it can be done—and quickly, if need be.

This is not the place to lay out a detailed global plan to cut carbon emissions, but a few examples of how to cut the use of oil and coal, the principal sources of carbon emissions, will illustrate the possibilities. One simple step that motorists can take to reduce oil use dramatically is to shift to cars with hybrid gas-electric engines. Automobiles such as the Toyota Prius and the hybrid Honda Civic that are already on the market are remarkably fuel-efficient. The 2004 Prius averages 55 miles per gallon in combined city and highway driving—double or even triple that of other midsize cars. If the United States were to raise the fuel efficiency of its automobile fleet over the next 10 years to that of today's Toyota Prius, U.S. gasoline consumption could be cut in half. This would not require any reduction in the number of cars used or in miles driven, only the use of more-efficient engines.[26]

But this is not the end. The hybrid gas-electric cars, which embody the most sophisticated automotive engineering on the road today, open up two exciting additional possibilities. The first is to modestly expand the electrical storage capacity of the hybrids by adding a second battery. The second is to include a plug-in recharge capacity so that owners can recharge their car batteries at night, when electricity demand drops, leaving surplus generating capacity. Given the typical U.S. daily commute of 12 miles roundtrip, these two steps would allow commuting and local driving, such as shopping, to be done almost entirely with electricity, saving gasoline for the occasional longer trip. Adding a second battery and a plug-in capacity could reduce gasoline use by perhaps another 20 percent, for a total reduction in U.S. gas use of 70 percent.[27]

These two modest technological modifications lead to an exciting possibility on the supply side, namely the use of cheap wind-generated electricity to power automobiles. Does the United States have the wind power potential to do this? As described later in this chapter, it has enough harnessable wind power to meet its electricity needs several times over.[28]

There are similarly exciting possibilities for cutting coal use. If, for example, the world were besieged by high temperatures and rising food prices, it would be a simple matter to replace the widely used old-fashioned, highly inefficient, incandescent light bulbs with compact fluorescent lamps that provide the same light but use less than a third as much electricity. A worldwide decision to phase out incandescent light bulbs would allow literally hundreds of coal-fired power plants to be closed. Not only would this help stabilize climate, but the return on investment in the new bulbs in the form of lower electricity bills is roughly 30 percent a year.[29]

These are but two of the obvious things that can be done on the demand side to cut carbon emissions. Reducing U.S. gasoline use for automobiles by 70 percent and dramatically cutting electricity use for lighting are exciting prospects for reducing dependence on imported oil, lowering the trade deficit, and stabilizing climate. We simply need a bit of imagination, some leadership, and a modest additional investment.

Turning to Renewable Energy Sources

There are also many options for cutting carbon emissions by harnessing renewable sources of energy, including wind energy, solar energy, geothermal energy, and biomass. Each of these can be developed in many ways. On the solar front, there are solar electric cells, solar thermal power plants, and the direct use of solar energy for water and space heating. The most immediately promising short-term source of new energy is wind. It is a vast resource, one that could meet all the world's electricity needs. As this chapter aims simply to give a sense of the possibilities for cutting carbon emissions, the discussion here will focus only on wind as a renewable source of energy.

The use of wind power is growing fast because it is abundant, cheap, inexhaustible, widely distributed, clean, and climate-benign—a set of attributes that no other energy source can match. Consider the U.S. potential. In 1991, the U.S. Department of Energy published a national wind-resource inventory. It concluded that North Dakota, Kansas, and Texas alone had enough harnessable wind energy to satisfy national electricity needs. For many people this was a surprise. They had no idea wind was such a vast resource.[30]

In retrospect, this was a gross underestimate because it was based on the wind energy that could be harnessed by the wind turbine technologies of 1991. Design advances since then enable turbines to operate at lower speeds, to convert wind into electricity more efficiently, and to harvest a much larger wind regime. Whereas the average wind turbine in 1991 might have been 40 meters tall, today turbines are closer to 100 meters, reaching heights where winds are stronger and steadier than they are at the earth's surface. While in 1991 the government concluded that wind power in just three states could satisfy national *electricity* needs, it may now be that these three states have enough harnessable wind energy to meet national *energy* needs. Although it is helpful to use these three wind-rich states to illustrate the scale of U.S. wind resources, many of the other 47 states are also richly endowed with wind energy.[31]

Europe is the model for developing wind power. Although its wind resources are modest compared with those of the United States, it is moving much faster to harness them. In its late 2003 projections, the European Wind Energy Association (EWEA) shows Europe's wind-generating capacity expanding from 28,400 megawatts in 2003 to 75,000 megawatts in 2010 and then 180,000 megawatts in 2020. By 2020, just 16 years from now, projections show that wind-generated electricity will be able to satisfy the residential needs of 195 million Europeans, half of the region's population.[32]

Europe is tapping its offshore wind resources as well as those on land. A 2004 assessment of Europe's offshore potential by the Garrad Hassan wind energy consulting group concluded that if governments move aggressively to develop their vast offshore resources, wind could be supplying all of the region's residential electricity by 2020.[33]

Wind-generating capacity worldwide, growing at over 30 percent a year, has jumped from less than 5,000 megawatts in 1995 to 39,000 megawatts in 2003—nearly an eightfold increase. (See Figure 7–3.) In comparison, natural gas use leads the fossil fuels, with an annual growth rate topping 2 percent during the same period, followed by oil at less than 2 percent and coal at less than 1 percent. Nuclear generating capacity expanded by 2 percent.[34]

The modern wind-generating industry was born in California during the early 1980s, but the United States, which now has 6,400 megawatts of generating capacity, has fallen behind Europe in adopting this promising new technology. Germany overtook the United States in 1997; within Europe, it leads the way with 14,600 megawatts of generating capacity. Spain, a rising wind power in southern Europe, may overtake the United States in 2004. Tiny Denmark, which led Europe into the wind era with the development of its own wind resources, now gets an impressive 20 percent of its electricity from wind. It is also the world's leading manufacturer and exporter of wind turbines.[35]

When the wind industry first began to develop in California, wind-generated electricity cost 38¢ per kilowatt-hour. Since then it has dropped to 4¢ or below in prime

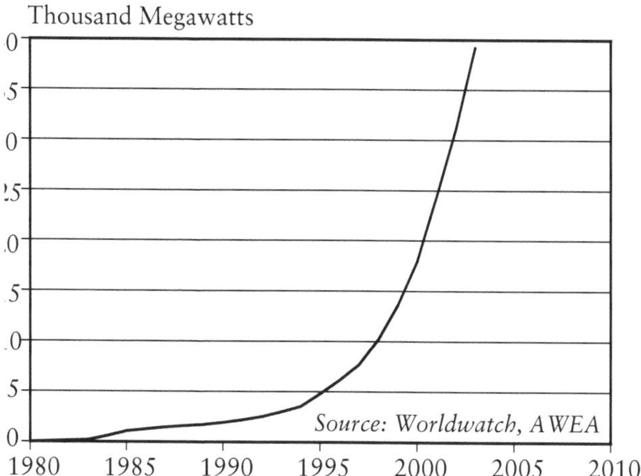

Figure 7–3. *World Wind Electric Generating Capacity, 1980–2003*

wind sites. And some long-term supply contracts have been signed for 3¢ per kilowatt-hour. EWEA projects that by 2020 many wind farms will be generating electricity at 2¢ per kilowatt-hour, making it cheaper than other sources of electricity.[36]

The United States is lagging in developing wind energy not because it cannot compete technologically with Europe in manufacturing wind turbines but because of a lack of leadership in Washington. The wind production tax credit of 1.5¢ per kilowatt-hour, which was adopted in 1992 to establish parity with subsidies to fossil fuel, has lapsed three times in the last five years—most recently at the end of 2003, when Congress failed to pass a new energy bill. The uncertainty about when it would be renewed disrupted planning throughout the wind power industry.[37]

The United States—with its advanced technology and wealth of wind resources—should be a leader in this field. Unfortunately the country continues to rely heavily on coal, a nineteenth-century energy source, for much of its electricity at a time when European countries are replacing coal with wind. Europe is not only leading the world into the wind age, it is also leading the world into the post-fossil-fuel age—the age of renewable energy and climate stabilization. By demonstrating the potential for harnessing the energy in wind, Europe is unveiling the new energy economy for the rest of the world.

The impetus for that new energy economy to unfold quickly may come from an unexpected source: agriculture. The effect of rising temperatures on crop yields fundamentally broadens the responsibility for food security. Historically, food security was the sole responsibility of the Ministry of Agriculture, but now the Ministry of Energy also bears responsibility. Decisions made by ministries of energy on whether to stay with carbon-based, climate-disrupting fossil fuels or to launch a crash program to develop renewables may have a greater effect on food security than do any of the decisions made in ministries of agriculture.

Data for figures and additional information can be found at www.earth-policy.org/Books/Out/index.htm.

8

Reversing China's Harvest Decline

The phenomenal rise in China's grain production from 90 million tons in 1950 to 392 million tons in 1998 was one of the great economic success stories of the late twentieth century. But in 1998 production peaked and turned downward, falling to 322 million tons in 2003. As noted in Chapter 1, this drop of 70 million tons exceeds the entire grain harvest of Canada. Thus any attempt to expand the world grain harvest enough to rebuild depleted world grain stocks starts with reversing the decline in China.[1]

Virtually all of China's production decline of nearly 18 percent from 1998 to 2003 is the result of a 16-percent shrinkage in grain area. Several forces are at work here, as described in Chapter 5. Cropland is being converted to nonfarm uses at a record rate, including industrial and residential construction and the paving of land for roads, highways, and parking lots. With deserts expanding by 360,000 hectares (1,400 square miles) a year, drifting sands are covering cropland in the north and west, making agriculture impossible. The loss of irrigation water is also reducing the harvested area, particularly of wheat, which is grown in the northern, drier regions of the country.[2]

In 2004 China's improved grain harvest, lifted by a

substantial rise in the rice support price and unusually favorable weather, was expected to regain 21 million of the 70-million-ton-drop of the preceding five years. Even with this projected production increase, China's harvest in 2004 will still fall short of consumption by 35 million tons. And there are several worrying trends that undermine the hope that the harvest will rise consistently again anytime soon.[3]

Grainland Shrinking

Chapter 1 described "the Japan syndrome," a set of interacting trends that explain why grain production declines in countries that are already densely populated before they industrialize. Each of the three countries discussed—Japan, South Korea, and Taiwan—had virtually identical experiences. In short, as industrialization gains momentum, grain consumption and grain production both rise, more or less together. In a relatively short time, however, grain planted area begins to shrink as farmland is converted to nonfarm uses, as grain is replaced by higher-value fruit and vegetable crops, and as the migration of farm labor to the cities reduces double cropping. This shrinkage in grain area then leads to declining grain production.[4]

China is facing precisely the same forces that within three decades cut grain harvests by one third to one half in Japan, South Korea, and Taiwan. But China's challenge is even greater because it is also losing grainland to expanding deserts and it is faced with spreading water shortages that are shrinking the grain harvest—problems the other three countries did not have.

China's deserts are advancing as its 1.3 billion people and 404 million cattle, sheep, and goats put unsustainable pressure on the land. Indeed, desert expansion has accelerated with each successive decade since 1950. The Gobi is marching eastward and is now within 150 miles of Beijing. Some deserts have expanded to the point where they are starting to merge. Satellite images show the Bardanjilin in north-central China pushing southward toward the Tengry desert to form a single, larger desert, overlapping Inner Mongolia and Gansu provinces. To the west in Xinjiang province, two much larger deserts—the Taklamakan and the Kumtag—are also heading for a merger.[5]

Wang Tao, Deputy Director of the Cold and Arid Regions Environmental and Engineering Research Institute, the world's premier desert research institute, reports that on average 156,000 hectares were converted to desert each year from 1950 until 1975. From 1975 to 1987, this increased to 210,000 hectares a year. But in the 1990s, it jumped to 360,000 hectares annually, more than doubling in one generation.[6]

The human toll is heavy, but rarely is it carefully measured. Wang Tao estimates that 24,000 villages "have been buried [by drifting sand], abandoned or endangered seriously by sandy desertification" affecting some 35 million people. In effect, Chinese civilization is retreating before the drifting sand that covers the land, forcing farmers and herders to leave. Most of this abandonment has come over the last two decades.[7]

Overplowing and overgrazing are converging to create a dust bowl of historic dimensions. With little vegetation remaining in parts of northern and western China, the strong winds of late winter and early spring can remove literally millions of tons of topsoil in a single day—soil that can take centuries to replace. For the outside world, it is dust storms like the ones described in the beginning of Chapter 5 that are drawing attention to the deserts forming in China.

The removal of small soil particles by wind in dust storms marks the early stages of desertification. This is followed by sand storms as desertification progresses. The growing number of major dust storms, as compiled by the China Meteorological Administration, indicates how rapidly this is happening. After increasing from 5 in the 1950s to 14 during the 1980s, the number leapt to 23 in the 1990s. (See Table 8–1.) The current decade began with more than 20 major dust storms in 2000 and 2001 alone.[8]

While overplowing is now being partly remedied by paying farmers to plant their grainland in trees, overgrazing continues largely unabated. China's cattle, sheep, and goat population tripled from 1950 to 2003. While the United States, a country with comparable grazing capacity, has 96 million cattle, China has a slightly larger herd of 103 million. But for sheep and goats, the figures are 8 million versus a staggering 317 million. Concentrated in the western and northern provinces of Inner Mongolia, Xingjiang, Qinghai, Tibet, and Gansu, sheep and goats are destroying the land's protective vegetation. The wind does the rest, removing the soil and converting grassland into desert.[9]

Even as overgrazing destroys forage, the number of sheep and goats continues to increase. While China's cat-

Table 8–1. *Number of Major Dust Storms in China, by Decade, 1950–99*

Decade	Number
1950–59	5
1960–69	8
1970–79	13
1980–89	14
1990–99	23

Source: See endnote 8.

tle herd has scarcely doubled since 1950, the number of sheep has nearly tripled and goat numbers have quintupled. The disproportionate growth of the goat population is a telltale sign of a deterioration in forage quality, a shift that favors the hardier goats.[10]

Millions of rural Chinese are being uprooted and forced to migrate eastward as the drifting sand covers their cropland. Expanding deserts are driving villagers from their homes in Gansu, Inner Mongolia, and Ningxia provinces. An Asian Development Bank assessment of desertification in Gansu Province reports that 4,000 villages risk being overrun by drifting sands.[11]

A report by a U.S. embassy official in May 2001 after a visit to Xilingol Prefecture in Inner Mongolia (Nei Mongol) notes that the prefecture's livestock population climbed from 2 million as recently as 1977, just before the economic reforms, to 18 million in 2000. With the economic reforms, the government lost control of livestock numbers. A Chinese scientist doing grassland research in the prefecture notes that if recent desertification trends continue, Xilingol will be uninhabitable in 15 years.[12]

The U.S. Dust Bowl of the 1930s forced some 2.5 million "Okies" and other refugees to leave the land, many of them heading from Oklahoma, Texas, and Kansas to California. But the dust bowl forming in China is much larger, and during the 1930s the U.S. population was only 150 million—compared with 1.3 billion in China today. Whereas the U.S. flow of Dust Bowl–refugees was measured in the millions, China's will be measured in the tens of millions.[13]

While the deserts are expanding, so too are the cities. With the fastest economic growth of any country since 1980, the voracious land hunger in the residential, industrial, and transportation sectors is consuming vast areas of land—much of it cropland. The sheer size of China's

Mohair goats, deprived of adequate forage as grasslands deteriorate, graze on each other. Herders wrap their goats in discarded clothes to protect them. Inset: A goat with all its hair eaten by other poorly nourished goats. Photo: Lu Tongjing.

population of 1.3 billion is impressive, but even more impressive is the fact that 1,193 million of them live in 46 percent of the country. The five sprawling provinces of Tibet, Qinghai, Xinjiang, Gansu, and Inner Mongolia, which account for 54 percent of the country's area, have only 81 million people—just 6 percent of the national total. (See Figure 8–1.) Thus industrial and residential construction and the land paved for roads, highways, and parking lots will be concentrated in less than half of the country, where 94 percent of the people live. People are crowded into this region simply because this is where the arable land and water are.[14]

Local government enthusiasm for establishing development zones for commercial and residential buildings or industrial parks in the hope of attracting investment and jobs is consuming cropland at a record pace. The Ministry of Land and Resources reported in early 2004 that some 6,000 development zones and industrial parks cover some 3.5 million hectares. In 2003, the Ministry of Land Resources reported the conversion of a record 2.1 percent of cropland to nonfarm uses, alarming political leaders in Beijing.[15]

Cars, as mentioned earlier, are also taking a toll. Every 20 cars added to China's automobile fleet require the paving of an estimated 0.4 hectares of land (1 acre, or roughly the area of a football field) for parking lots, streets, and highways. Thus the 2 million new cars sold in 2003 meant paving over 40,000 hectares of land—the equivalent of 100,000 football fields. If this was cropland, and most of it probably was, it could have produced 160,000 tons of grain—enough to feed half a million Chinese.[16]

If China had Japan's automobile ownership rate of one car for every two people, it would have a fleet of 640 million, a fortyfold increase from the 16 million cars of

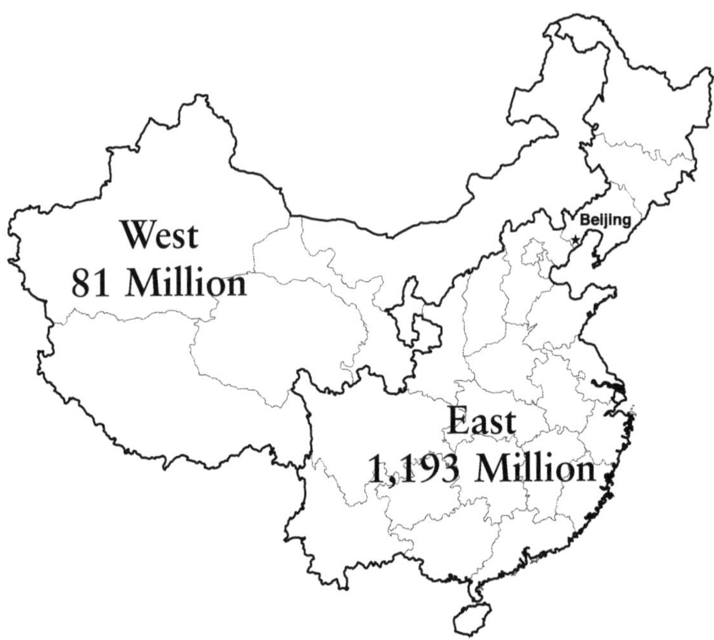

Figure 8–1. *Geographic Distribution of Population in China, 2002*

today. Such a fleet would require paving over almost 13 million hectares of land—again, most of it likely cropland. This figure is equal to nearly two thirds of China's 21 million hectares of riceland—land that produces 120 million tons of rice, the country's principal food staple. When farmers in southern China lose a hectare of double-cropped riceland to the automobile, rice production takes a double hit.[17]

Farmers throughout China are also converting grainland to higher-value harvests. In a country where farms average 0.6 hectares (1.6 acres), the only readily available option for boosting income for many is to shift to higher-value crops. In each of the last 11 years, the area in fruits and vegetables has increased, expanding by an average of 1.3 million hectares per year. This jump in area from 10 million hectares in 1991 to 26 million hectares in 2003 (see Figure 8–2) included hefty increases in asparagus, cabbage, carrots, cauliflower, peppers, eggplant, garlic, onions, spinach, watermelons, tomatoes, apples, pears, and grapes. The huge expansion is in response to rapid growth both in domestic demand (as incomes rise and diets diversify) and in the export market. High-value, labor-intensive export crops are well suited to a country where labor is by far the most abundant resource.[18]

In the more prosperous coastal provinces, the migration of farmworkers to cities has made it more difficult to double crop land. For example, the once widespread practice of planting wheat in the winter and corn as a summer crop depends on quickly harvesting the wheat as soon as it ripens in early summer and immediately preparing the seedbed to plant the corn. But with mil-

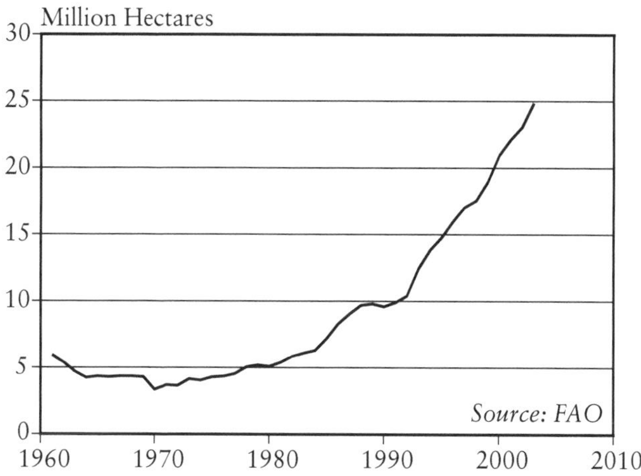

Figure 8–2. *Vegetable and Fruit Harvested Area in China, 1961–2003*

lions of younger workers moving to cities in search of jobs, many villages no longer have enough able-bodied workers to make this quick transition, and the double-cropped area is reduced.

Reversing the fall in grain production will not be easy simply because growth in the activities that are claiming cropland is so relentless. Reversing any one of these trends—conversion to nonfarm uses, desert expansion, the decline in multiple cropping—will take an enormous effort. While higher grain prices may temporarily increase multiple cropping and boost production, China faces an uphill battle in sustaining growth in its grain harvest for the same reasons that Japan, South Korea, and Taiwan did.[19]

The new economic incentives introduced by Beijing in early 2004 to boost grain production may modify some of these trends in the short run. For example, higher support prices for rice and wheat may slow the movement of rural labor into cities. They also encourage farmers to invest more in inputs, such as fertilizer and pest control. And if adoption of the new incentives coincides with unusually favorable weather, a modest, short-term upturn in grain production can occur, as it did in 2004. But restoring sustained growth in the grain harvest will challenge the leadership in Beijing.[20]

An Aquacultural Initiative

One of China's responses to land and water shortages has been to vigorously expand aquacultural output, taking advantage of this grain-efficient form of animal protein. Although fish farming goes back some 3,000 years in China, annual aquacultural output did not reach 1 million tons until 1981, shortly after the 1978 economic reforms. It then began to expand rapidly, climbing from 1 million tons in 1981 to 28 million tons in 2002.[21]

China's fast-growing aquacultural sector totally dominates world aquaculture. Indeed, as of 2002, China produced 28 million tons out of the world aquacultural output of 40 million tons, accounting for more than two thirds of the global total.[22]

Within China, the area used for aquaculture production, both fresh water and offshore, totals some 6.8 million hectares—roughly the size of Ireland or West Virginia. This area consists of farm-built ponds; reservoirs, including many smaller ones used for water storage; and the offshore areas occupied by cages. China has some 800,000 cages used for fish production that are near offshore.[23]

Carp dominate China's output, at nearly 13 million tons—almost half of the 28-million-ton annual harvest. Other freshwater fish, including tilapia, push the freshwater finfish total to 15 million tons. There are also more than 5 million tons of oysters, mussels, and scallops produced, including both freshwater and saltwater species. The remaining 8 million tons consists of a variety of species, including crab, prawn, and eel.[24]

As China's aquacultural output has grown, it has spawned a huge aquafeed industry, totaling 16 million tons in 2003—11 million tons of grain and 5 million tons of soybean meal. Freshwater fish rations are now roughly one third soybean meal, substantially higher than the 18–20 percent soymeal content in livestock and poultry feeds. Traditionally, fish feeds relied heavily on fish meal to achieve the optimum protein content, but with fish becoming scarce, soybean meal has proved to be a readily acceptable substitute for China's largely omnivorous fish species.[25]

Consumption of farmed fish per person is easily two times higher in cities than in the countryside. Because cities are dispersed, so too is fish farming. Most of the fish are produced by small farmers who typically build their own ponds or use local reservoirs.[26]

The extraordinary growth in China's aquacultural output is largely the result of strong government support for the industry. China is also exporting substantial quantities of aquacultural products. The U.S. agricultural attaché's office in Beijing reports that China exports some $2 billion of aquatic product a year to Japan. Other leading markets include the United States, South Korea, Hong Kong, and Germany, with totals ranging from $1 billion to the United States to $185 million to Germany. A recent U.N. Food and Agriculture Organization study projects that China's aquatic product consumption will rise by 80 percent over the next five years.[27]

Water Shortages Spreading

Throughout the northern half of China water tables are falling, wells are going dry, and rivers are being drained dry before they reach the sea. The irrigation water prospect in the North China Plain, which produces half of China's wheat and a third of its corn, is one of the keys to China's long-term food security.[28]

Farmers in this region rely on three rivers and two aquifers for irrigation water. The three rivers in the region, from north to south, are the Hai, Huang (Yellow), and Huai. The North China Plain has two aquifers, one shallow and one deep.[29]

The Yellow River, the second largest river in China after the Yangtze, is often referred to as the cradle of Chinese civilization. Originating on the Tibetan Plateau, it flows through eight provinces en route to the sea. Unfortunately, in many recent years it has been drained dry, failing to reach the sea during the dry season.[30]

The Hai river basin, the northernmost of the three, includes two of China's largest cities—Beijing and Tianjin, with 14 million and 11 million people, respectively. The whole basin, which contains 100 million people, is now in chronic deficit. The Sandia National Laboratory, which has modeled the water balance in China's rivers, concluded that water withdrawals in the Hai River basin of 55 billion tons in 2000 exceeded the sustainable supply of 34 billion tons by 21 billion tons. This deficit is made up by groundwater mining. When the aquifer is depleted, the water supply in the basin will drop sharply.[31]

Urbanization is directly affecting the water balance in the Hai River basin. When villagers migrate to cities, where they have indoor plumbing, water consumption typically multiplies fourfold. Finding jobs in industry for the millions of new workers moving into the region imposes additional demands on the dwindling water supply. With competition for water between farmers, cities, and industry intensifying, irrigated agriculture in the Hai River basin may largely disappear by 2010.[32]

Demands on Huai river water, the southernmost of the three rivers, comes from both Anhui and Jiangsu provinces. Like the other two rivers, it also is sometimes drained dry, failing to make it to the sea. Originating in the mountains to the immediate west of the North China Plain, the Huai is a key source of water for farmers in both Anhui and Jiangsu provinces.[33]

The North China Plain depends heavily on two aquifers—a shallow aquifer that is replenishable and a deep fossil aquifer, which is not replenishable. Farmers, cities, and industries are pumping from both. Where the shallow aquifer has been depleted, the amount of water pumped is necessarily reduced to the amount that is recharged.[34]

In many areas now, the deep aquifer is the principal source of water, but it too is being depleted. When this finally happens, pumping will come to an end. He Qingcheng, head of the groundwater monitoring team in the Geological Environmental Monitoring Institute, observes that with depletion of the deep aquifer, the region is losing its last water reserve—its only safety cushion.[35]

Water shortages will shape the evolution of China's economy in fundamental ways. The gravity of the water situation in the North China Plain can be seen in the frenzy of well drilling in recent years. At the end of 1996, the five provinces of the North China Plain—Heibei, Henan, Shandong, and the city provinces of Beijing and Tianjin—had 3.6 million wells, the bulk of them for irrigation. A detailed study of the situation in 1997 showed 99,900 wells abandoned as they ran dry. Partly to compensate, some 221,900 new wells were drilled. The desperate quest for water in China is evident as well drillers go to ever greater depths, often using technology borrowed from the oil drilling industry.[36]

Concerns about the tightening water situation are reflected in a World Bank report: "Anecdotal evidence suggests that deep wells [drilled] around Beijing now have to reach 1,000 meters (more than half a mile) to tap fresh water, adding dramatically to the cost of supply."[37]

Turning Abroad for Grain

Each of the grains that together account for 96 percent of China's production—wheat, rice, and corn—is suffering a decline. Even with an improved wheat harvest in 2004, production still fell short of consumption by 12 million tons, an amount equal to the entire wheat harvest of Argentina. When the country's wheat stocks are depleted within the next year or so, the entire shortfall will have to be covered from imports. In some ways, the rice deficit is even more serious. Trying to cover an annual rice shortfall of 10 million tons in a world where annual rice exports total only 26 million tons could create chaos in the world rice economy. And with a corn shortfall of 12 million tons and stocks already largely depleted, China may soon be importing corn as well.[38]

Before the 70-million-ton drop in China's grain production from 1998 to 2003, the country was producing a modest exportable surplus of 5–10 million tons a year. (See Figure 8–3.) Now this has changed. By 2003, grain production had fallen 56 million tons below consumption. With the harvest upturn in 2004, the shortfall improved but still stood at 35 million tons.[39]

China has been covering its grain shortfall in recent years by drawing down its stocks. After peaking at 326 million tons in 1999, China's carryover stocks of grain plummeted to 102 million tons in 2004. (See Figure 8–4.) At this level, stocks amount to little more than pipeline supplies and cannot be drawn down much farther. This means that within another year or two shortfalls will have to be covered entirely by importing grain.[40]

The decline in the grain harvest from 1998 to 2003 alarmed China's leaders. So did the rise in grain prices

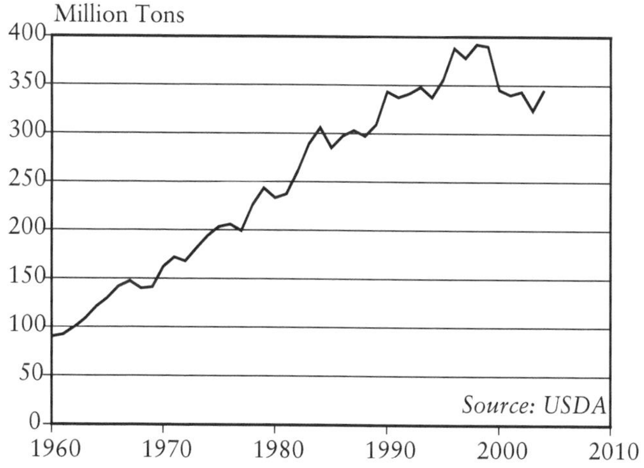

Figure 8–3. *Grain Production in China, 1960–2004*

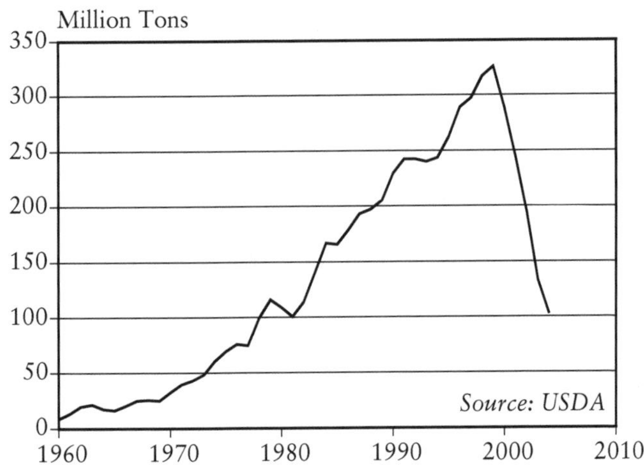

Figure 8–4. *Grain Stocks in China, 1960–2004*

beginning in the fall of 2003. The year-to-year rise of nearly 30 percent in grain prices between 2003 and 2004 forced the government to draw down its shrinking stocks of grain even faster in an effort to stabilize food prices.[41]

In late 2003 and early 2004, Chinese wheat-buying delegations purchased 8 million tons of wheat in Australia, the United States, and Canada. Within two years China went from being essentially self-sufficient to being the world's leading wheat importer. In March China made small purchases of rice from Thailand and Viet Nam for immediate import, suggesting that the internal rice situation, at least in some localities, was also beginning to tighten. In late August 2004, Beijing sought to buy 500,000 tons of rice from Hanoi, but was told that, given the export restrictions designed to ensure domestic rice price stability, Viet Nam could not deliver any rice until early 2005.[42]

Concerned with falling production and the threat of

politically destabilizing rises in food prices, the government announced an emergency appropriation in March 2004—increasing its agricultural budget by 20 percent or roughly $3.6 billion. The additional funds were to be used to raise support prices for wheat and rice, the principal food staples, and to improve irrigation infrastructure. For the State Council to approve such an increase outside the normal budgeting process indicates the government's mounting concern about food security. Nearly all the leaders in Beijing today are survivors of the great famine of 1959–61, when 30 million Chinese starved to death. For them, food security is not an abstraction.[43]

On March 29, 2004, the government announced that the support price for the early rice crop would be raised by 21 percent. This got farmers' attention, as Beijing hoped it would, leading them to plant nearly 2 million additional hectares of rice—a gain of 7 percent from 2003. China's rice harvest rose from 112 million tons in 2003 to an estimated 126 million tons in 2004. This 14-million-ton gain was the result of both stronger incentives and a recovery from last year's weather-depressed yields. Overall, grain production was up 21 million tons in 2004. The much smaller gains for wheat and corn were, as with rice, due to a combination of better weather and stronger prices.[44]

While stronger prices can temporarily reverse the decline in China's grain production, they do not eliminate the forces that are shrinking China's grainland area and thus its harvest. Unless Beijing can quickly adopt policies to protect its cropland, continued shrinkage of the grain harvest and mounting dependence on imported grain may be inevitable.

A sense of how quickly China can turn to the world market can be seen with soybeans. As recently as 1997, the nation was essentially self-sufficient in soybeans. (See

Figure 8–5.) In 2004, it imported 22 million tons—dwarfing the 5 million tons imported by Japan, formerly the world's leading soybean importer. The Chinese economy is so large and so dynamic that its import needs can shake the entire world. Its soaring soybean needs, combined with a sub-par harvest in the United States in 2003, led to a temporary doubling of world soybean prices during the early months of 2004.[45]

Over the longer term, China's grain imports are likely to climb to levels never before seen. Japan, South Korea, and Taiwan today each import roughly 70 percent of their total grain supply. If China were to do the same, it would be importing 280 million tons per year. This exceeds current world grain imports by all countries of just over 200 million tons. This is obviously not going to happen, but what sort of adjustments will prevent China from following the path of Japan, South Korea, and Taiwan? What sort of economic stresses will develop in

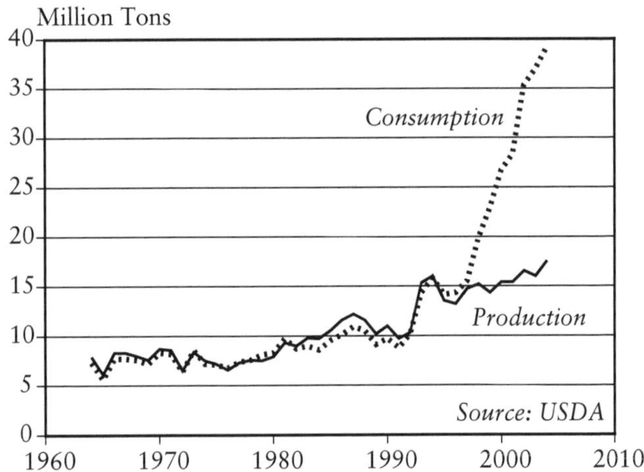

Figure 8–5. *Soybean Production and Consumption in China, 1964–2004*

the world as China willingly or unwillingly is pushed in the same direction as the earlier Japan syndrome countries? What sort of stresses will develop within China if the world cannot supply the vast imports it needs?[46]

A New Food Strategy

The freefall in China's grain production from 1998 to 2003 indicates what can happen if Beijing continues with business-as-usual on the farm front. If China is to avoid a long-term decline in its grain harvest, it will need radical new policies and a basic reordering of priorities in the national budget. Future food security depends on policy shifts in land ownership, water pricing, desert reclamation, and transportation.[47]

Following the economic reforms of 1978, the huge farm production teams were dissolved and the "Family Responsibility System" was introduced. Individual farm families were leased a plot of land for a 15-year term. When these began expiring in the 1990s, they were replaced with 30-year leases. For farmers, having their own plot of land to farm unleashed an enormous burst of energy in the countryside, one that boosted grain production from 199 million tons in 1977 to 306 million tons in 1984.[48]

Unfortunately, even with these long-term leases farmers are still insecure because the land can be taken from them at any time by local officials. Arthur Kroeber writes in the *Financial Times* that China's "village leaders can arbitrarily 'readjust' land rights at a moment's notice, changing boundaries or even forcing farmers to move from an old plot to a new one." They can also confiscate a farmer's land and sell it for industrial development projects, compensating farmers at far below market value. There is no recourse because the farmers do not have title to the land. They are tenants, not owners. The authority of village leaders to appropriate land at will is thus a threat hanging over the heads of villagers, a source of political control.[49]

If tenants become owners, however, production might surge again. Giving farmers title to their land could harness latent energies in the countryside, encouraging them to invest in land improvements that yield long-term productivity gains, such as terracing and local water storage facilities. Taking this next step would help rejuvenate China's sagging agriculture, but it would also mean that local party officials would lose control of the land, and with it a large measure of political power.

Another key to reversing the decline in China's grain production is to accelerate its program to raise water productivity, particularly in the northern half of the country, where water shortages are strangling agriculture. This means pricing water at a level that reflects its value in a water-scarce situation. Higher prices combined with economic incentives to shift to more water-efficient technologies, whether in irrigation, in industry, or at the household level, can expand output while reducing water use to where water tables can be stabilized.

China also needs a reliable system of grain price supports that will encourage farmers to invest more in agriculture. They need not be particularly high, but they do need to be reliable. In 1994, when China raised support prices by 40 percent, it generated a strong production response, but then prices were permitted to gradually decline to the world market level over the next several years. With prices so low that farmers were no longer earning a profit, many of them simply lost interest and produced only enough grain for their own needs. Without reliable price supports that enable farmers to grow grain profitably, China's food security is at risk. The decision in early 2004 to raise the rice support price by 21 percent was a step in the right direction.[50]

One of the innovative responses to China's growing demand for animal protein is the development of the world's most advanced aquacultural sector. At the heart of this effort is the highly efficient carp polyculture pioneered by the Chinese and described in Chapter 3, which enabled Chinese fish farmers to produce more than 15 million tons of freshwater fish in 2002. For China, this emphasis on the highly efficient production of animal protein is another positive step—and an example for other countries to follow.[51]

China, facing the growing competition between cars and crops for land, may soon be forced to reexamine its transportation policy. There is an inherent conflict between continuing to build an auto-centered transportation system and ensuring future food security. With nearly 1.2 billion of its 1.3 billion people living in less than half of the country on the eastern and southern coast, the competition between cars and farmers for land will be intense. If China moves toward having a car in every garage, American-style, it will face not only gridlock but soaring food shortfalls as well. If Beijing continues to expand the production and ownership of automobiles, cropland will almost certainly continue to shrink. The alternative is to develop a passenger transport system centered on high-tech rail and buses, augmented by bicycles. Such a system would provide not only more mobility in the end, but also greater food security.[52]

In some ways the most intractable environmental problem China faces is the growth of deserts throughout the western and northern parts of the country largely as a result of overgrazing. Unless the central government makes a concerted effort to reduce the population of sheep and goats to the carrying capacity of the grazing lands, deserts will continue their eastward march toward Beijing and the blinding dust storms that mark the late winter and early spring will become even more frequent.

Planting marginal cropland in trees helps correct the mistakes of overplowing, but it does not deal with the overgrazing issue. Arresting desertification may depend more on grass than trees—on both enabling existing grasses to recover and planting grass in denuded areas.

Beijing is trying to arrest the spread of deserts by asking pastoralists to reduce their flocks of sheep and goats by 40 percent, but in communities where wealth is measured in livestock numbers and where most families are living under the poverty line, such cuts are not easy. Some local governments are requiring stall-feeding of livestock with forage gathered by hand, hoping that confinement of herds will permit grasslands to recover.[53]

China is taking some of the right steps to halt the advancing desert, but it has a long way to go to reduce livestock numbers to a sustainable level. At this point, there is not yet a plan in place that will halt the advancing deserts. Qu Geping, the farsighted Chairman of the Environment and Resources Committee of the National People's Congress, estimates that the remediation of land in the areas where it is technically feasible would cost $28.3 billion. Halting the advancing deserts will thus require a massive commitment of financial and human resources, one that may force a choice between the large investments proposed for south-north water diversion projects and those required to halt the advancing deserts that are occupying more of China each year.[54]

China is faced with an extraordinary challenge. Adopting the needed policies in agriculture, water, land ownership, desert reclamation, and transportation to ensure future food security will be far more demanding than for countries that developed earlier, when land and water were more plentiful. Stated otherwise, if China is to restore and sustain a rise in grain production, it will have to adopt measures in land use planning, transportation, and water use that are responsive to its unique circumstances—measures that no government has ever adopted. The entire world has a stake in China's success.

Data for figures and additional information can be found at www.earth-policy.org/Books/Out/index.htm.

9

The Brazilian Dilemma

We may be facing a seismic shift in the geography of world food trade as China emerges as a massive food importer and Brazil becomes a leading food exporter. While China is losing cropland rapidly, Brazil is gaining it at a record rate, setting the stage for a fast-expanding agricultural link between the two countries.

Over the last few decades, the dominant bilateral food-trade link was between the United States, the leading exporter of grain, soybeans, and meat, and Japan, the top importer of these commodities. Signs that the Brazil-China link could eclipse the U.S.-Japan link are already in evidence with soybeans. China is now the world leader, importing some 22 million tons in 2004—more than four times Japan's soy imports of 5 million tons. Meanwhile, Brazil replaced the United States as top exporter, shipping 44 million tons of soybeans, including soybean meal and oil, to other countries in 2004 compared with 33 million tons from the United States.[1]

In 2004, China also displaced Japan as the world's number one wheat importer. It may soon do the same for feedgrains. If Brazil can accelerate the growth in its grain harvest to match that of soybeans over the last decade, it will have a large exportable surplus of grain to help cover the expanding needs of importing countries such as China. However, it will be exceedingly difficult for Brazil to duplicate the soybean expansion for both economic and ecological reasons.[2]

There are also signs that China may move ahead of Japan as an importer of meat in the not-too-distant future. In some recent years, China has imported more poultry than Japan has. With imports of pork rising, China may overtake Japan here as well. With beef, however, Japan's imports lead the world, while China's are still negligible. On the export side, Brazil's fast-growing exports of pork, poultry, and beef are in the process of overtaking those of the United States. Barring some unexpected event, Brazil soon will be the world leader.[3]

The pressures to push back the agricultural frontier in Brazil will be intense in the next few decades, since this is the only country with a vast land area that potentially can be cropped. Economic forces and political pressures for Brazil to expand its cultivated area are strong and growing stronger. The world urgently needs more grain and high-quality protein. Projections indicate that nearly 3 billion people will be added to world population by 2050, some 5 billion people in developing countries want to move up the food chain, 840 million people are still chronically hungry and malnourished, and the backlog of technology to raise land productivity is shrinking. Throughout the late twentieth century, additional demand for food from population growth translated into efforts to raise land productivity, but now as that becomes more difficult, continuing population growth is generating pressure to expand the cultivated area.[4]

This pressure to clear more land means the worst fears of environmentalists may be realized. The prospect of losing so much of the earth's remaining biological diversity is scary, to say the least. In our increasingly integrated world, the fate of both Brazil's Amazon basin and the *cerrado*—a savannah-like region the size of Europe on the basin's southern edge—can no longer be separated from the family planning decisions of hundreds of millions of couples outside of Brazil and the aspirations for a better diet of billions more.

Can Brazil dramatically expand its cropland area and avoid the ecological catastrophe that followed on the heels of the last major cropland expansion initiative, the Soviet Virgin Lands Project in the 1950s? Can Brazilian agriculture expand in a way that will respond to growing world food needs and at the same time protect the rich diversity of life in the Amazonian rainforest and the *cerrado*?[5]

World's Leading Source of Soybeans

For Brazil, the door into the soybean world opened in 1972 with the collapse of the massive Peruvian anchovy fishery, a leading worldwide source of protein supplements in livestock and poultry rations. Since this fishery accounted for one fifth of the world fish catch and for an even larger share of animal feed protein supplements before its demise, its abrupt collapse created a protein shortage that drove soybean prices off the chart. These steep price rises, combined with a U.S. soybean export embargo in 1973 when Washington tried to check the inflationary rise in domestic food and feed prices, set the stage for Brazil's entry into the market. The embargo, which raised concerns about the reliability of the United States as a supplier, led importing countries in Europe plus Japan to encourage soybean production in Brazil and Argentina.[6]

In a prescient move, the Brazilian government invested heavily in a comprehensive soybean research program, including the breeding of varieties adapted specifically to local soils and growing conditions throughout the country. Government leaders also started thinking seriously about how to create the infrastructure needed to link the country's vast unplowed interior to world markets. These research initiatives, along with economic incentives, boosted Brazil's soybean production from 1 million tons in 1969 to 15 million tons in 1980.[7]

Initially, production growth was concentrated in the traditional farming regions in the south—the states of Rio Grande do Sul, Santa Catarina, Paraná, and São Paulo—but after 1990 it began to spread rapidly into the *cerrado*. (See Figure 9–1.) *Cerrado* soils are highly acidic, saturated with aluminum, and low in phosphorus, with a

Shaded area is the cerrado

Figure 9–1. *The Cerrado of Brazil*

limited capacity to store water. These characteristics provided a formidable barrier to cultivation until Brazilian scientists discovered that adding 3–8 tons of lime per hectare reduced the acidity and neutralized the free aluminum in the soil. Once this was done, the deep well-drained soils of this savannah-like region could be farmed. Liming and heavy fertilization, combined with the breeding of varieties that could tolerate higher aluminum levels, set the stage for the expansion.[8]

On the downside, as Kenneth Cassman of the University of Nebraska notes, it is likely that soil organic matter will deteriorate rapidly in these tropical and subtropical soils, where temperature, humidity, and abundant rainfall all favor the decomposition of organic matter and crop residues. This contrasts with the U.S. Corn Belt, where cold winters slow down soil decomposition. The carbon sequestration on this land once it has been tilled for a few years will be far less than in the original *cerrado*, thus contributing to higher atmospheric carbon dioxide levels.[9]

Analysts estimate that the Brazilian *cerrado* includes an additional 75 million hectares (185 million acres) of potentially cultivable land, an area almost as large as the U.S. area planted to grain and soybeans. Although Brazil now produces one third of the world's soybeans, U.S. Department of Agriculture experts believe the country has the potential to easily triple its current soybean production.[10]

Argentina has also achieved hefty gains in soybean production, but its potential for continuing rapid expansion is limited compared with Brazil. Indeed, part of Argentina's soybean expansion has been at the expense of grain.[11]

Brazil's soybean production has expanded at a pace rarely matched for a major crop in any country. In 1969, Brazil was producing only 1 million tons of soybeans. (See Figure 9–2.) By 1986, it produced 13 million tons

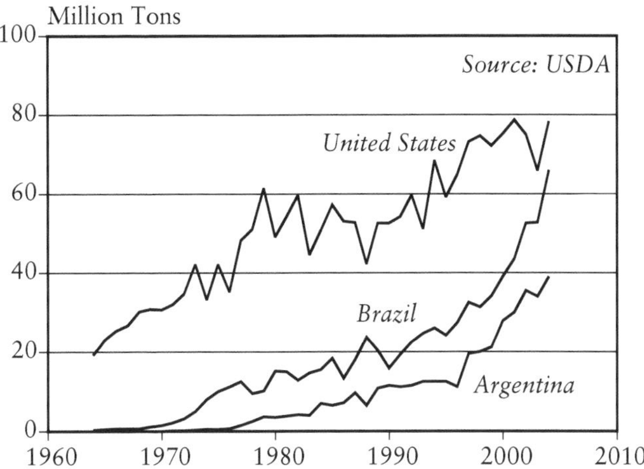

Figure 9–2. *Soybean Production by Country, 1964–2004*

compared with a U.S. output of 53 million tons. Since then, Brazil's production has expanded more than five-fold—jumping to 66 million tons in 2004, compared with U.S. production of 78 million tons. Within the next few years, Brazil is likely to become the world's leading soybean producer, a position held by the United States since it displaced China a half-century ago. While Brazil can expand soybean output severalfold, the U.S. potential for expansion is limited by the lack of new land to plow.[12]

On the import side of the equation, China's soybean imports, which were negligible a decade ago, are now four times those of Japan, the traditional leader. (See Figure 9–3.) For several decades the largest movement of soybeans between two countries was that between the United States and Japan. Now the largest bilateral flow is between Brazil and China.[13]

By 2004, Brazil's 24 million hectares of soybeans had exceeded not only its area of corn, wheat, and rice individually, but the area of all of them combined. The 2004

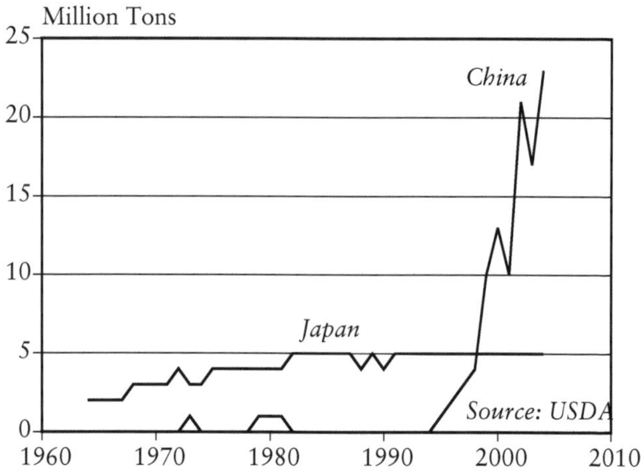

Figure 9–3. *Soybean Imports into China and Japan, 1964–2004*

soybean harvest of 66 million tons exceeded the grain harvest of 60 million tons (Figure 9–4), marking the first time an oilseed harvest has eclipsed that of grain in any large agricultural country. In the United States, the area in soybeans passed that in wheat in 1978 and now rivals that planted to corn. Even so, total U.S. soybean production of 78 million tons in 2004 is scarcely one fifth the size of the U.S. grain harvest of 360 million tons.[14]

Brazil's national agricultural research network, EMBRAPA, has worked hard and successfully to adapt temperate-zone soybean varieties to Brazil's subtropical growing conditions. Reflecting its success, the soybean yield per hectare in Brazil today has edged above that in the United States, long the world leader.15

Despite Brazil's extraordinary successes, future expansion will not always be easy. Brazil's soybean growers are faced with a debilitating Asian rust disease that is now curbing yields. Spraying crops with a fungicide to

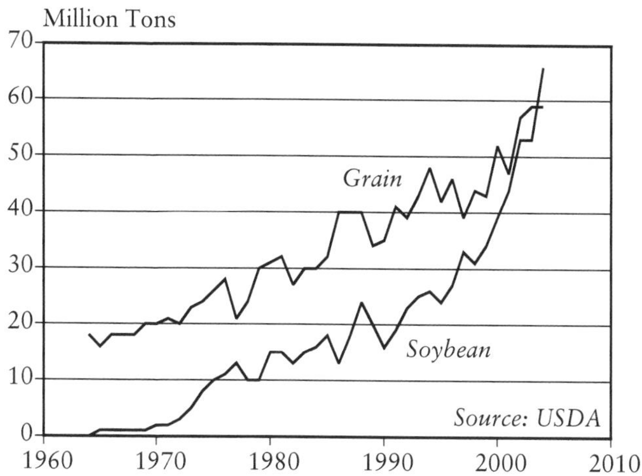

Figure 9–4. *Grain and Soybean Production in Brazil, 1964–2004*

control the disease, which cost $1.2 billion in 2003 and 2004, is sometimes ineffective because frequent rains wash the treatment off the plants. In some areas, the cost of protecting the soybeans from this damaging disease may now go as high as 50¢ per bushel, which represents roughly 8 percent of the crop's value based on prices over the last decade.[16]

A lack of infrastructure, principally roads and electricity, also hinders Brazil's soybean expansion. Because the *cerrado* on average is some 1,600 kilometers by road from the east coast ports, getting soybeans from the interior to points of export is costly. Although Brazil's cheap land gives its soybean farmers a huge edge over their U.S. counterparts, the United States has a well-developed barge system for moving the product from the Midwest down the Mississippi River to the port of New Orleans. Beans destined for Asia also can move easily by rail from the U.S. Midwest to West Coast ports such as Seattle and Portland.[17]

Transportation costs loom large for Brazilian exports of soybeans and grain. First the commodities must be moved to a port either on the coast or on one of the tributaries of the Amazon. Ships loading at Amazon ports have to travel more than 1,500 kilometers, or nearly a thousand miles, merely to get to the Atlantic Ocean. If they are going to Asia, they must then either go north through the Panama Canal or south around the Cape of Good Hope. Either way, the distance is some 20,000–22,500 kilometers. Even shipping to Europe is costly. Marty McVey and his colleagues with AGRI Industries point out that shipping soybeans from Sapezal, Mato Grosso, to Europe costs $1.59 a bushel, while from Iowa it is only 84¢, barely half as much.[18]

Within Brazil, simply getting soybeans from the more remote parts of Mato Grosso, which straddles the *cerrado* and the Amazon basin, to port can be costly. In a world where oil prices are likely to be rising, the variation in transportation costs of moving soybeans, corn, or meat to the outside world could shape Brazil's pattern of development, pushing it toward meat exports rather than the far bulkier shipments of grain.[19]

Creating the agriculture transport infrastructure within Brazil will take time and, among other things, vast amounts of capital investment. Nonetheless, these barriers are not insurmountable. Soybean output will likely continue expanding until Brazil becomes the world's largest producer, most likely well before the end of this decade.

Feed Supplier for the World?

Brazil's impressive capacity to raise soybean production has raised questions about whether it could also become a leading world supplier of grain for food and feed. As of 2004, the country is a modest net importer of grain and has been for several decades. Like other tropical countries, it has difficulty producing wheat in its tropical and subtropical regions. Brazil's wheat is produced almost entirely in its southernmost states, on the border with Argentina. Given the heavy fertilizer requirements in the *cerrado*, wheat production costs in the expansion region are nearly double those of Argentina and the United States. It thus seems unlikely that Brazil will emerge as a wheat exporter unless world wheat prices climb far beyond current levels.[20]

Wheat and rice are Brazil's two food staples. The nation consumes roughly 10 million tons of wheat per year, producing half and importing half. In contrast, it consumes roughly 8 million tons of rice a year and is essentially self-sufficient. Given the tightening rice situation in Asia, could Brazil boost its rice production enough to produce a surplus for export to Asia? Is there enough water in Brazil's rice-growing states, all in the south, to expand production of this water-intensive grain? The Amazon basin has an abundance of water, but are its soils suited to rice production?[21]

Corn, which totally dominates Brazil's grain harvest at over 40 million tons a year, is used primarily as a feedgrain. Until recently, Brazil imported corn, but it is now self-sufficient and typically exports a few million tons per year. Corn rotates well with soybeans since the latter fix nitrogen, for which corn has a ravenous appetite. Soybeans grown in rotation with corn are less vulnerable to damage from disease and insects, but corn and soybeans are not a perfect marriage in Brazil, simply because corn yields are relatively low on the *cerrado* soils. While Brazil's soybean yields match or exceed those in the United States, its corn yields run around 3.5 tons per hectare, compared with U.S. yields of 9 tons. In addition, corn grown on the nutrient-poor *cerrado* soils requires heavy doses of fertilizer, especially nitrogen. Unfortunately, nitrogen leaches through these porous soils, leading to high nitrogen levels in both surface and underground water.[22]

Beyond these agronomic and environmental issues, transport costs are formidable. Although a bushel of corn is worth less than half as much as a bushel of soybeans in the world market, the cost of transporting it from the remote interior to the coastal ports is the same. Whether Brazil could overcome the combination of heavy fertilizer requirements, low yields, and high transportation costs to become a major corn exporter remains to be seen.[23]

Corn is not the only feedgrain option. Sorghum is also a possibility. Although sorghum output in Brazil is limited, the annual harvest has jumped from less than a million tons to over 2 million tons within the last three years. Since this is a drought-tolerant crop that does well during the dry season, it could find an ecological niche in the rotation systems in the drier regions of the Brazilian *cerrado*.[24]

Brazil's annual net grain imports of 8 million tons during the 1990s have dropped to a modest 3 million tons, mostly wheat, during the current decade. Given the robust character of the country's agriculture, the net imports could be eliminated entirely and Brazil could become at least a small net exporter, largely on the strength of its corn exports. The key question is, How much would the world corn price have to rise to justify a large expansion in production for the world market?[25]

Brazil has demonstrated clearly that when the world soybean price is $6 a bushel or above, farmers will invest in the land clearing and the government will invest in the needed infrastructure to expand soybean production and exports rapidly. It is doubtful, however, that it can produce large quantities of corn for the world market at the $2.50 per bushel price of recent years if the transport cost to Europe is $1.59 a bushel, as it is for soybeans. It does not seem likely that Brazil will become a major supplier of grain to the world unless corn prices rise to $4 or so. Brazil's weakness as a grain producer is evident when it is compared with the United States. While it is about to overtake the United States in soybean production, it produces only 60 million tons of grain compared with 360 million tons in the United States. (See Figure 9–5.)[26]

Meat Exports Climbing

An expanding world market for meat combined with surging domestic consumption is spurring rapid growth in Brazil's beef, pork, and poultry sectors. Total meat exports have expanded from scarcely a half-million tons in 1990 to 4 million tons in 2004, enabling Brazil to challenge the United States for world leadership.[27]

Brazil has the world's largest commercial cattle herd,

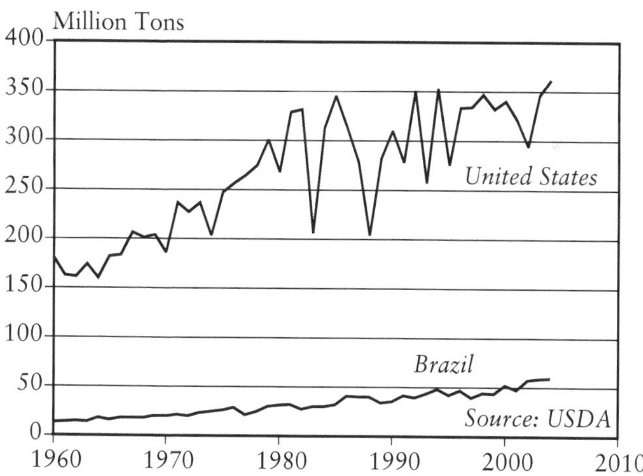

Figure 9–5. *Grain Production in Brazil and the United States, 1960–2004*

with 190 million cattle. (See Figure 9–6.) With the eradication of foot-and-mouth disease in the key cattle-raising states—including Mato Grosso, Rondônia, and Tocatins, which straddle the *cerrado* and the Amazon—and with eradication of this disease expected nationwide by 2005, many new markets have opened up for Brazilian beef. Interested buyers include not only industrial countries, such as those in Western Europe, but also developing countries, such as Chile, Egypt, and Saudi Arabia. In anticipation of this export growth, the annual increase in the Brazilian herd jumped from less than 2 million during the 1990s to 6 million from 2000 to 2004. Much of this growth is occurring on the edge of the Amazon.[28]

Brazilian beef exports leapt from 200,000 tons in 1995 to 1.4 million tons in 2004, just edging out Australia and the United States, the traditional leaders in beef exports. Growth in demand for beef was driven by the expanding domestic market until the December 1998 devaluation of the Brazilian reál, which made Brazilian

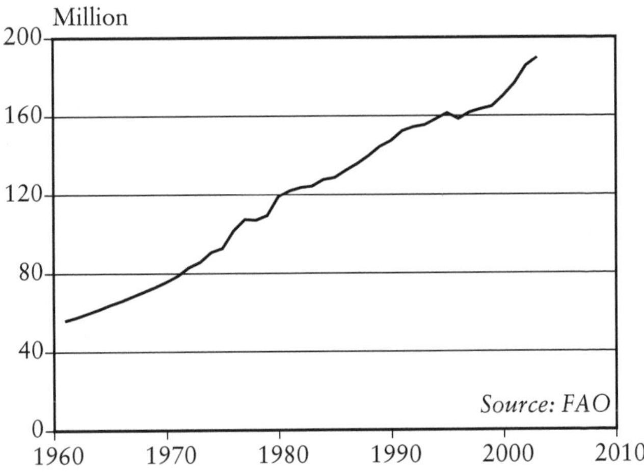

Figure 9–6. *Number of Cattle in Brazil, 1961–2003*

beef much more competitive in world markets. (See Figure 9–7.) The resulting expansion of exports raised the price of beef in the Amazon.[29]

In contrast to the situation with beef, Brazil is a second-tier producer of pork, with only 2.5 million tons a year compared with 9 million tons in the United States and a staggering 46 million tons in China. Yet Brazilian exports of 400,000 tons of pork put it third among exporting countries, trailing only Canada and the United States.[30]

With poultry, Brazil is both a leading producer and exporter. Its rapidly growing production may overtake China within the next few years, leaving it second only to the United States. Exports have climbed to 2.2 million tons in 2004, matching those of the United States.[31]

In summary, Brazil's exports of beef, pork, and poultry are expanding steadily. It is the leading exporter of beef, ranks third in pork, and is vying with the United States for the lead in poultry. With beef, Brazil is essentially exporting grass, part of it grown on land in the

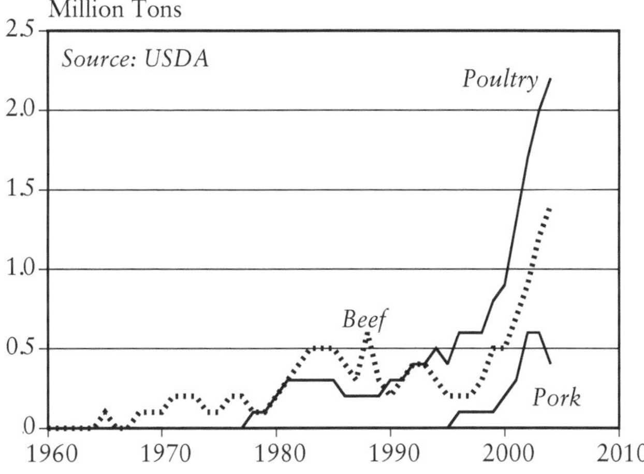

Figure 9–7. *Meat Exports from Brazil, 1960–2004*

Amazon basin that was until recently covered by rainforest. But when it comes to pork and poultry, it is basically grain that is being exported. While no precise data are available, Brazil appears to be exporting close to 10 million tons of grain in the form of meat. It may be that the country's future lies not so much in exporting grain per se as it does in exporting grain indirectly in the form of pork and poultry.[32]

Domestic Demand Growing

Brazil's capacity to export grain will be affected by its fast-growing domestic needs, fueled by a population that is currently expanding by 2 million per year. By 2050, Brazil's population is projected to reach 233 million, four fifths of the current U.S. population of 297 million. Annual income per person is projected to go from $2,400 today to $26,000 in 2050. This compares with $27,000 in Canada today and $34,000 in the United States.[33]

As incomes rise, Brazilians will move up the food chain, consuming more grain-intensive livestock products. Although meat consumption is dominated by beef, which is almost entirely grass-fed, the consumption of pork and poultry, both grain-fed, is rising. At present, two thirds of the grain used in Brazil is consumed indirectly in the form of livestock products. Of the nearly 44 million tons of animal feed used in 2003, 24 million tons were for poultry and egg production, 13 million for pork, 4 million for cattle, and 3 million for other uses. Grain used for feed is likely to continue rising in parallel with the consumption and export of grain-rich livestock products.[34]

One way of assessing future demand for grain is to look at recent trends. Between 2000 and 2004, annual grain consumption in Brazil grew 2 million tons a year. If it continues to grow this much on average as both its population size and income levels approach those of the

United States today, total grain consumption would climb to 154 million tons by 2050. This compares with current U.S. grain consumption of roughly 240 million tons per year and would mean that Brazil's farmers would need to add roughly 100 million tons of grain production to current output just to remain self-sufficient.[35]

Stated otherwise, Brazil would need to nearly triple its grain harvest by 2050 to satisfy the growth in domestic demand. By comparison, over the last half-century the United States has doubled its grain harvest, nearly all of it by raising land productivity. Given this large projected domestic demand for grain and the high cost of shipping corn to markets in Europe and Asia, Brazil will not easily develop a large exportable surplus of grain.[36]

Expansion: The Risks and Costs

Brazil has embarked on a massive expansion of its cropland area. Unlike the land planted to grain, which has changed little over the last three decades, staying at around 20 million hectares a year, the area in soybeans exploded from 1 million hectares in 1970 to 24 million hectares in 2004. Half of this growth came after 1996, most of it in the *cerrado*, with the remainder in the Amazon basin.[37]

But is this expansion sustainable? As noted earlier, the last massive cropland expansion anywhere in the world was the Soviet Union's Virgin Lands Project during 1954–60. Within a matter of years, the Soviets had plowed an area of grassland for wheat production that was larger than the wheatlands of Canada and Australia combined. Although it boosted production initially, this plan soon turned into an ecological disaster.[38]

Not long after the expansion, which was centered in Kazakhstan, a huge dust bowl began to form. Not only has half the land now been abandoned, but the wheat yield on the remaining land is scarcely 1 ton per hectare—one sixth the yield in Western Europe.[39]

Many ecologists are concerned about soil erosion in the *cerrado* if this region is cleared of vegetation on the scale that now seems likely. In the state of Mato Grosso there is already evidence of damaging wind erosion. To the west, across the border in Bolivia, soil erosion is undermining land productivity in an area near Santa Cruz that pioneered growing soybeans beginning in 1970.[40]

One advantage that the *cerrado* has over the land cleared in the Soviet Union is that rainfall is much heavier, typically 39–75 inches a year. This helps explain why the yield per hectare of Brazil's soybeans, grown largely in the *cerrado*, has eclipsed that of the United States, the traditional leader.[41]

While the clearing of the *cerrado* is largely soybean-driven, that of the Amazon is much more cattle-driven. Nevertheless, it is the vast worldwide market for soybeans that is financing the transport infrastructure in Brazil's interior, both within the *cerrado* and into the neighboring Amazon. This is what makes the Amazon accessible to small farmers, commercial farmers, and cattle ranchers. Phillip Fearnside, a leading authority on environmental issues in Brazil says, "Soybeans are much more damaging than other crops, because they justify massive transportation infrastructure projects that unleash a chain of events leading to the destruction of natural habitats over wide areas in addition to what is directly cultivated for soybeans."[42]

Beyond this, the commercial strength of soybean production also enables growers to buy land that already has been cleared by cattle ranchers and by small farmers located either near or in the Amazon, which drives the sellers further into the Amazon in their quest for cheaper land. Thus while the soybean is an unrivaled source of protein in a protein-hungry world, it is also a powerful new threat to the biological diversity of Brazil.[43]

Unfortunately, the Brazilian government itself is working to open the Amazon to development. The principal umbrella for this, a program known as Avança Brasil (Advance Brazil), is intended to open up areas to industrial, agricultural, logging, and mining activities in a way that will accelerate development of the Brazilian economy. A recent article in *Science* reports, "Investment totaling about $40 billion over the years 2000–2007 will be used for new highways, railroads, gas lines, hydroelectric projects, power lines, and river channelization projects. The Amazonian road network is being greatly expanded and upgraded, with many unpaved sections being converted to paved, all weather highways."[44]

The devaluation of the reál and the progressive eradication of foot-and-mouth disease together have raised the price of beef and the profitability of cattle ranching in the Amazon. It is accelerating expansion "of the region's road and electricity network and large investments in modern slaughterhouses and meat-packing and dairy plants," according to the Center for International Forestry Research. The center further notes, "Very low land prices in the Amazon also help to make ranching profitable. These prices remain very low in part because farmers find it easy to illegally occupy government land without being prosecuted, and to deforest areas much larger than the 20 percent of each landholding currently permitted by law."[45]

As roads are cut through the Amazon, pulling settlers, loggers, and ranchers further into the region, the forest is becoming increasingly fragmented. Once the rainforest canopy is disrupted, the incoming sunlight dries out the land, leaving the understory vegetation vulnerable to fire. As a result, fires that are intentionally set to clear land sometimes burn out of control, making the forest more vulnerable to fires caused by lightning. A healthy rainforest does not burn simply because it is too wet, but once the forest is fragmented, it dries out and loses this natural defense.

One of the principal manifestations of this vulnerability is the growing number of forest fires now systematically recorded by satellites. The fire season in the Amazon, now an annual occurrence, has become an identifiable phenomenon only in the last few decades.[46]

In addition to the soil erosion and degradation associated with the loss of forest cover, there is a risk that the forest clearing could jeopardize the recycling of rainfall inland. The traditional agricultural region in the Brazilian south, not to mention in neighboring Bolivia, Paraguay, Uruguay, and northern Argentina, is watered by moisture-laden air masses from the Atlantic that move westward across the Amazon and then flow south as they approach the Andes.[47]

As land is cleared of vegetation for either farming or cattle ranching, its capacity to recycle rainfall inland is reduced. Some 20 years ago, Brazilian scientists Eneas Salati and Peter Vose published a landmark article in *Science* analyzing the effect of deforestation on rainfall recycling in the Amazon. They noted that when rain from the moisture-laden air masses that originated over the Atlantic fell on healthy rainforest, about one fourth ran off, returning to the Atlantic Ocean, and three fourths evaporated into the atmosphere either directly or through transpiration and was then carried further inland to again come down as rainfall. This explains how rainforests get their name. It also explains why rainfall is heavy throughout the Amazon basin and south of it, in the *cerrado*, as well.[48]

By contrast, Salati and Vose showed that when rain falls on land that is cleared for grazing or cropping, the runoff/evaporation ratio is reversed as roughly three fourths of it runs off, returning to the sea, leaving only one fourth to evaporate and be carried further inland. Thus the loss of at least 2 million hectares of Amazonian forest a year is slowly weakening the water recycling mechanism that brings water to the agricultural regions of south-central Brazil.[49]

Another cost, not only for Brazil but for the world, of clearing vast areas of the Amazon rainforest and the *cerrado* to produce corn and soybeans and to graze cattle is the loss of plant and animal species. The Amazonian rainforest is one of the most biologically rich regions in the world. Although there are mechanisms in place that are designed to protect this diversity, such as the requirement that landowners clear no more than one fifth of their land, the government lacks the capacity to enforce this.[50]

The *cerrado* is also biologically rich, with thousands of endemic plant and animal species. It contains many large mammals, including the maned wolf, giant armadillo, giant anteater, deer, and several large cats—jaguar, puma, ocelot, and jaguarundi. The *cerrado* contains 837 species of birds, including the rhea, a cousin of the ostrich, which grows six feet tall. More than 1,000 species of butterflies have been identified. Conservation International reports that the *cerrado* also contains some 10,000 plant species—at least 4,400 of which are not found anywhere else.[51]

On March 15, 2004, President Lula da Silva announced an "action plan to prevent and control deforestation in the Legal Amazon." This plan allocates $135 million to a range of activities, including land use planning and greater enforcement of laws concerning both the illegal occupation of government lands and their deforestation. It also commits resources to monitor deforestation using satellite images. Notwithstanding this and other similar initiatives in the past, the forces that are driving the growing world demand for soybeans and beef, which in turn are driving the deforestation, continue to gain momentum.[52]

According to Brazil's National Institute of Space Research, just over 2.5 million hectares of the forest in the Amazon disappeared in 2002. If anything, that number is likely to increase when the data for 2003 become available. From 1990 to 2000, cumulative deforestation in the Brazilian Amazon increased from 42 million hectares to 59 million, an average of 1.7 million hectares per year. The area of Amazonian forest lost over the decade was equal to two Portugals.[53]

A recent article in *Science* summed up the situation: "Conserving Amazonian forests will not be easy. If the world expects Brazil to follow a development path that differs from its current one, and from a path that most developed nations have followed in the past, then substantial costs will be involved. The investment, however, would surely be worth it. At stake is the fate of the greatest tropical rainforest on earth."[54]

If there is no coordinated effort for developing Brazil's interior, including both the *cerrado* and the Amazon, that integrates economic and environmental goals, many species will be threatened and countless numbers of them will likely disappear. This could lead to the greatest single loss of plant and animal species in history, biologically impoverishing not only Brazil but the planet on a scale that we cannot easily imagine.

Data for figures and additional information can be found at www.earth-policy.org/Books/Out/index.htm.

10

Redefining Security

In each of the first four years of this new century, world grain production has fallen short of consumption. The shortfalls in 2002 and 2003, the largest on record, and the smaller ones in 2000 and 2001 were covered by drawing down stocks. These four consecutive shortfalls in the world grain harvest have dropped stocks to their lowest level in 30 years. When there are no longer any stocks to draw down, the only option will be to reduce consumption.[1]

In early 2004, world grain prices were up some 20 percent over previous years. Soybean prices were double the levels of a year earlier. The combination of stronger prices at planting time and the best weather in a decade raised the 2004 grain harvest by 124 million tons to 1,965 million tons, up 7 percent. For the first time in five years, production matched consumption, but only barely. Even with this exceptional harvest, the world was still unable to rebuild depleted grain stocks.[2]

The immediate question is, Will the 2005 harvest be sufficient to meet growing world demand, or will it again fall short? If the latter, then world grain stocks will drop to their lowest level ever—and the world will be in uncharted territory on the food front.

The risk is that another large shortfall could drive

prices off the top of the chart, leading to widespread political instability in low-income countries that import part of their grain. Such political instability could disrupt global economic progress, forcing world leaders to recognize that they can no longer neglect the population and environmental trends that have created harvest shortfalls in four out of the last five years. While terrorism will no doubt remain an important policy issue, the threat posed by growing food insecurity may dwarf it in terms of the number of lives lost and the extent of economic disruption.

The Tightening Food Supply

The world food supply is tightening because world grain demand is continuing to expand at a robust pace while production growth is slowing as the backlog of unused agricultural technology shrinks, cropland is converted to nonfarm uses, rising temperatures shrink harvests, aquifers are depleted, and irrigation water is diverted to cities.

The world's population is projected to increase by nearly 3 billion by 2050. Two thirds of this growth will occur in the Indian subcontinent and in Africa, the world's hungriest regions. Most of the other 1 billion people will be born in the Middle East, which faces a doubling of its population, and in Latin America, Southeast Asia, and the United States. This projected population increase requires more land not only to produce food but also for living space—homes, factories, offices, schools, and roads.[3]

Some countries are still expanding their cropland, including, for example, Indonesia and Malaysia, both of which are converting rainforest into oil palm plantations. Yet in these two countries, the area of land being cleared is quite small compared with what could happen in Brazil. As noted in Chapter 9, the remaining potential for expanding the world's cultivated area is concentrated in this large South American country. But when this expansion potential is set against heavy cropland losses elsewhere to residential and industrial construction, to the paving of land for automobiles, and to the spread of deserts, the potential net growth in world cropland is likely to be modest at best.[4]

In many countries the irrigation water supply is shrinking as aquifers are depleted. But even as wells are going dry, irrigation water is being diverted to fast-growing cities. Farmers are getting a smaller share of a diminishing supply. Perhaps even more important, recent research indicates that higher temperatures reduce grain harvests, and at a time when we face the prospect of continually rising temperatures.[5]

In contrast to the last half-century, when the world fish catch quintupled to reach 93 million tons, we cannot expect any growth in the fish catch at all during the next half-century. The growing world demand for seafood must now be satisfied entirely from aquaculture, where fish are fed mostly grain and soymeal. This puts additional pressure on the earth's land and water resources.[6]

Beyond these various environmental and resource trends that are affecting the food prospect, the world's farmers are now also wrestling with a shrinking backlog of agricultural technology. For the world's more progressive farmers, there are few, if any, unused technologies that will substantially raise land productivity. Even more serious, dramatic new yield-raising technologies are likely to be few and far between.[7]

We can also look at the world food prospect through the lens of the Japan syndrome, the sequence of events that occurs in countries that are densely populated before they industrialize. The changes that led to the peaking of grain production and its subsequent decline in Japan,

South Korea, and Taiwan seem certain to affect many other countries. China is the first major one to experience a precipitous decline in its grain harvest. In 1995, when I projected in *Who Will Feed China?* that China's grain harvest would drop, I sensed that this was imminent. But when the downturn came after 1998, it fell faster than I had expected.[8]

As we look at other large, densely populated countries, like India, we know that the same forces are at work, but we do not know exactly when the grain harvest will peak and begin to decline. It could be several years away. But that the preconditions for a decline exist there can be little doubt. Already India has a population density nine times that of the United States. The living space required for the 18 million people added each year to India's population of 1.1 billion means less and less land is available to produce food. Other countries, such as Indonesia, Bangladesh, Pakistan, Egypt, Nigeria, and Mexico, may also soon experience the Japan syndrome as modernization takes land from agriculture.[9]

The Politics of Food Scarcity

For more than 40 years, international trade negotiations have been dominated by grain-exporting countries—principally the United States, Canada, Argentina, and Australia—pressing for greater access to markets in importing countries. Now the world may be moving into a period dominated not by surpluses but by shortages. In this case, the issue becomes not exporters' access to markets but importers' access to supplies.[10]

The behavior of exporters in recent years shows why grain-importing countries should be concerned. In early September of 2002, Canada—following a harvest decimated by heat and drought—announced that it would export no more wheat until the next harvest. Two months later, Australia, another key exporter, said that because of a short harvest it would supply wheat only to its traditional buyers. And in the summer of 2003, during the crop-withering heat waves in Europe, the European Union announced that it would not issue any grain export permits until the supply situation improved.[11]

A similar situation developed in Russia following a poor harvest in 2003. Facing a rise in bread prices of more than 20 percent, in January 2004 the government imposed an export tax of 24 euros ($30) per ton on wheat, effectively ending wheat exports. The tax continued into May.[12]

In late August 2004, China approached Viet Nam to buy 500,000 tons of rice. The leaders in Hanoi responded by saying that the rice could not be supplied until the first quarter of 2005 at the earliest. This is because the Vietnamese government had imposed export limits of 3.5 million tons for the year, or just under 300,000 tons per month, out of fear that growing external demand for its rice would lead to overexporting and thus to rising domestic prices.[13]

This response is of interest because Viet Nam is the world's second-ranking rice exporter after Thailand. Thailand, Viet Nam, and the United States account for 16 million of the 25 million tons of world exports. In addition to China, more than 30 other countries import substantial amounts of rice, ranging from 100,000 tons a year for Colombia and Sri Lanka to 1.8 million tons for Indonesia.[14]

China's rice crop shortfall in 2004 of 10 million tons hangs over the world market like a sword of Damocles. Where the rice will come from is not clear. What China may do in trying to import such large quantities is simply drive up world rice prices. If it had attempted to cover its shortfall in 2004 entirely by imports, world rice prices would almost certainly have doubled or tripled, as they did in 1972–74, forcing low-income rice-consumers to tighten their belts.[15]

The big test of the international community's capacity to manage scarcity may come when China turns to the world market for massive imports of 30, 40, or 50 million tons of grain per year—demand on a scale that could quickly overwhelm world grain markets. When this happens, China will have to look to the United States, which controls nearly half the world's grain exports.[16]

This will pose a fascinating geopolitical situation: 1.3 billion Chinese consumers, who have a $120-billion trade surplus with the United States—enough to buy the entire U.S. grain harvest twice—will be competing with Americans for U.S. grain, driving up food prices. In such a situation 30 years ago, the United States would simply have restricted exports, but today it has a stake in a politically stable China. The Chinese economy is not only the engine powering the Asian economy, it is also the only large economy worldwide that has maintained a full head of steam in recent years.[17]

Within the next few years, the United States may be loading one or two ships a day with grain for China. This long line of ships stretching across the Pacific, like an umbilical cord providing nourishment, may link the two economies much more closely than ever before. Managing this flow of grain so as to satisfy the needs of consumers in both countries may become one of the leading foreign policy challenges of this new century.

The risk is that China's entry into the world market will drive grain prices so high that many low-income developing countries will not be able to import enough grain. This in turn could lead to political instability on a scale that will disrupt global economic progress. What began with the neglect of environmental trends that are impairing efforts to expand food production could translate into political instability on a scale that interferes with international trade and capital flows, thus halting economic progress. At this point, it will be clear that our economic future depends on addressing long-neglected environmental trends.

How exporting countries make room for China's vast needs in their export allocations will help determine how the world addresses the stresses associated with outgrowing the earth. How low-income, importing countries fare in this competition for grain will also tell us something about future political stability. And, finally, how the United States responds to China's growing demands for grain even as it drives up grain and food prices for U.S. consumers will tell us much about the shape of the new world order.

If substantially higher grain prices are needed to bring additional agricultural resources into play, whether in boosting water productivity, which effectively expands the supply, or bringing new land into play in Brazil, how will the world adjust? It may be that the laissez-faire, independent decisionmaking of national governments will have to blend into a more coordinated approach to managing food supplies in a time of scarcity.

Unfortunately, the government of China contributes to global food insecurity by refusing to release data on its grain stocks, leaving the international community to try and estimate them independently. This leads to a great deal of uncertainty and confusion, as can be seen in three substantial revisions of estimates for China's grain stocks in the last four years by the U.S. Department of Agriculture (USDA) and the Food and Agriculture Organization (FAO). While holding this information close to the vest gives Chinese grain buyers an advantage in the world market, it makes it extremely difficult for the world to plan for, and thus respond to, potentially huge future import needs.[18]

Because the last half-century has been dominated by excessive production and market surpluses, the world has had little experience in dealing with the politics of scarcity outside a brief period in 1972–74. In 1972, with a poor domestic harvest in prospect, the Soviets entered the world wheat market secretively and managed to tie up almost all the world's exportable supplies of wheat before governments of either exporting or importing countries realized what was happening. When this was followed by bad weather and below-average harvests over the next two years, world wheat and rice prices doubled, creating serious problems for importers. The United States, which accounted for half of the world's grain exports, restricted sales to certain countries; even food aid shipments favored "friendly countries"—those that supported the United States in U.N. votes, for example.[19]

As surpluses are replaced with scarcity, there is a need to pay more attention to carryover grain stocks, the amount in the bin when the new harvest begins. FAO has established a minimum-security stocks target of 70 days of consumption. Once stocks drop below this level, grain prices become volatile, often driven by the latest weather forecast in a key food-producing country. With stocks at 63 days of consumption in 2004, a major shortfall in one or two major food-producing regions in 2005 could create chaos in world grain markets. Yet despite the importance of this issue, it gets little attention at the U.N. Security Council. Nor has the G-8, the group of eight large industrial countries, focused on it in annual meetings.[20]

One reason food shortages do not get the attention they once did is because famine, in effect, has been redefined. At one time, famine was a geographic phenomenon. When a country or region had a poor harvest, its people often faced famine. Given the growing integration of the world grain economy and today's capacity to move grain around the world, famine is concentrated much less in specific geographic regions and much more among income groups. Food shortages now translate into higher worldwide prices that affect low-income people throughout the world, forcing many to try to tighten their belts when there are no more notches left.

In the event of life-threatening grain price rises, a tax on livestock products could help alleviate temporary shortages. This would reduce consumption of grain-fed livestock products—meat, milk, and eggs—and thus free up for human consumption a small share of the grain normally fed to livestock and poultry. As noted earlier, in the United States, a reduction in grain consumption per person from 800 to 700 kilograms by moving down the food chain somewhat would not only leave most Americans healthier, it would also reduce grain consumption by some 30 million tons. That would be enough to feed 150 million people in low-income countries. At a time when grain stocks are at an all-time low and the risk of dramatic price jumps is higher than at any time in a generation, a tax on livestock products is the one safety cushion that could be used to buy time to stabilize population and restore economic stability in the world food economy.[21]

Stabilizing the Resource Base

Future food security depends on stabilizing four key agricultural resources: cropland, water, rangeland, and the earth's climate system. Stabilizing the farmland base means protecting it from both soil erosion and the conversion to nonfarm uses. In China, for example, where the grain harvested area fell from 90 million hectares in 1999 to 77 million hectares in 2004, arresting the shrinkage depends on halting the expansion of deserts and controlling the conversion of grainland to nonfarm uses.[22]

Protecting water resources means stabilizing water tables. The overdrafting that lowers water tables also raises the energy used for pumping. For example, in some states in India half of all electricity is used for water pumping. Higher pumping costs ultimately mean higher food production costs.[23]

Protecting rangeland is an integral part of the food security formula not only because damage to rangeland from overgrazing reduces the livestock carrying capacity, but also because the dust storms that follow devegetation of the land can disrupt economic activity hundreds of miles away. The drifting sand that follows the conversion of rangeland to desert can also invade farming areas, rendering cultivation impossible.

Most important, we need to stabilize the climate system. Agriculture as we know it has evolved over 11,000 years of rather remarkable climate stability. The negative effect of higher temperatures on grain yields underlines the importance of stabilizing climate as quickly as possible.[24]

Stabilizing any one of these resources is demanding, but our generation faces the need to do all four at the same time. This is a demanding undertaking in terms of leadership time and energy and also in financial terms. As noted earlier, desert remediation in China alone will require estimated expenditures of some $28 billion.[25]

It may seem obvious that if water tables start to fall and wells begin to go dry, alarm bells would ring and governments would launch an immediate effort to reduce pumping and bring demand into balance with supply by adopting water conservation measures. But not one of the scores of countries where water levels are falling has succeeded in stabilizing its water tables.

Protecting the world's grainland is equally difficult. Advancing deserts are a formidable threat in countries such as Mexico, Nigeria, Algeria, Iran, Kazakhstan, India, and China. If governments continue to treat the symptoms of desertification and fail to address the root causes, such as continuing population growth and excessive livestock numbers, the deserts will continue to advance.[26]

Shielding cropland from nonfarm demands can also be politically complex. The cropland-consuming trends that are an integral part of the modernization process, such as building roads, housing, and factories, are difficult to arrest, much less reverse. And yet the world as a whole cannot continue indefinitely to lose cropland without eventually facing serious trouble on the food front.

Just understanding the complex of issues we are facing on the food security front is difficult. Fashioning an effective response and then implementing it is even more so. This is, in a sense, an enormous educational challenge because it requires national political leaders to master these difficult issues. If they do not, there is little chance that we will arrest the accelerating deterioration in agriculture's natural support systems and prevent the economic decline that eventually will follow.

A Complex Challenge

When grain harvests fell short, stocks declined, and prices rose during the last half of the twentieth century, there was a standard response. At the official level, the U.S. government would return to production part or all of the cropland idled under its commodity set-aside programs. At the same time, higher prices would encourage farmers worldwide to use more fertilizer, drill more irrigation wells, and invest in other yield-enhancing measures. Production would jump and shortages would disappear.[27]

Now the possible responses to shortages are more

demanding. First, the U.S. cropland set-aside program was dismantled in 1996, depriving the world of this long-standing backup reserve for world grain stocks. As of 2004, only the European Union is holding cropland out of use to limit production, but it is a small area, perhaps 3 million hectares. The United States does have some 14 million hectares (35 million acres) of cropland, much of it highly erodible, in the Conservation Reserve Program under 10-year contracts with farmers, nearly all planted in grass. In an emergency, part of it could be plowed and planted in grain, but it is mostly low-rainfall, low-yielding land in the Great Plains that would expand the U.S. harvest only marginally.[28]

The world today faces a situation far different from that of half a century ago. Diminishing returns are setting in on several fronts, including the quality of new land that can be brought under the plow, the production response to additional fertilizer applications, the opportunity for drilling new irrigation wells, and the potential of research investments to produce technologies that will boost production dramatically.

In 1950, opportunities for expanding the cultivated area were already limited, but there were still some to be found here and there. Together they helped expand the world grainland area by roughly one fifth. Today, in contrast, the only country that has the potential to increase the world grainland area measurably is Brazil. And doing this would raise numerous environmental questions, ranging from soil erosion to decreased carbon sequestration in the plowed areas.[29]

A half-century ago, every country in the world could anticipate using much more fertilizer. Today, using more fertilizer has little effect on production in many countries. And a half-century ago, the use of underground water for irrigation was almost nonexistent. Vast aquifers were waiting to be tapped, yielding a sustainable supply of irrigation water. Today, drilling more irrigation wells is likely only to hasten the depletion of aquifers and a resulting drop in food production.

Diminishing returns also affect agricultural research. Fifty years ago agricultural scientists were just beginning to adapt the high-yielding dwarf wheats and rices and the hybrid corn to widely varying growing conditions around the world. Today the plant breeding focus has shifted from raising yields to using biotechnology to develop varieties that are insect-resistant or herbicide-tolerant. Plant breeding advances may still raise yields 5 percent here or perhaps 15 percent there, but the potential for dramatic gains appears limited.[30]

The world has changed in other ways. As world population and the global economy expanded dramatically over the last half-century, the world quietly moved into a new era, one in which the economy began pressing against the earth's natural limits. In this new situation, activities in one economic sector can affect another. Historically, for example, what happened in the transport sector had little effect on agriculture. But in a world with 6.3 billion people, most of whom would like to own a car, auto-centered transport systems will consume a vast area of cropland.[31]

In the societies that first turned to cars as the principal means of transportation, there was no need for the transportation minister to consult with the agriculture minister. During the earlier development of the United States, for example, there was more than enough land for crops and cars. Indeed, throughout much of this era farmers were paid to hold land out of production. Now that has changed. Food security is directly affected by transportation policy today.

If densely populated countries like China and India

turn to cars as the primary means of transportation, they will pit affluent automobile owners against low-income food consumers in the competition for land. These nations simply do not have enough land to support hundreds of millions of cars and to feed their people.

The competition between cars and people for resources does not stop here. Some key food-producing countries, including the United States, are producing ethanol from grain for automotive fuel. In 2004, the United States used some 30 million tons of its 278-million-ton corn harvest to manufacture ethanol for cars. This tonnage, requiring nearly 4 million hectares (10 million acres) to produce, would be enough to feed 100 million people at average world consumption levels. Other countries building grain-fed ethanol plants include Canada and China. The competition between affluent motorists and low-income food consumers is thus not only for the land used to produce food, but also for the food itself.[32]

The other side of this coin is that if grain prices rise sharply, ethanol plants are likely to close, as they did in 1996 when grain prices went up temporarily. This would free up grain for food or feed, thus providing an additional buffer when world grain supplies tighten.[33]

The loss of momentum on the food front in recent years argues for reassessing the global population trajectory. Indeed, population policymakers may hold the key to achieving a humane balance between population and food. We can no longer take population projections as a given. The world cannot afford for any women to be without family planning advice and contraceptives. Today, however, an estimated 137 million women want to limit the size of their families but lack access to the family planning services needed to do so. Eradicating hunger depends on filling the family planning gap and creating the social conditions that will accelerate the shift to smaller families.[34]

Food security is affected not only by the food-population equation, but also by the water-population equation and the efforts of water resource ministries to raise water productivity. Indeed, since 70 percent of world water use is for irrigation, eradicating hunger may now depend on a global full-court press to raise water productivity. Everyone knows it takes water to produce food, but we often do not realize how water-intensive food production is and how quickly water shortages can translate into food shortages. The ministry of health and family planning needs to cooperate not only with the ministry of agriculture but also with the ministry of water resources. Those living in land-hungry, water-short countries need to know how their childbearing decisions will affect the next generation's access to water and to food.[35]

It is perhaps indicative of the complexity of the times in which we live that decisions on energy development made in ministries of energy may have a greater effect on the earth's temperature, and hence future food security, than decisions made in ministries of agriculture. Even so, ministers of energy are rarely involved in food security planning.

Ensuring future food security therefore can no longer be left to ministries of agriculture alone. Food security is now directly dependent on policy decisions in the ministries of health and family planning, water resources, transportation, and energy. This dependence of food security on an integrated effort by several departments of government is new. And because it has emerged so quickly, governments are lagging far behind in their efforts to coordinate these departments and their agendas.

One of the essentials for success in this new situation is strong national political leaders. In the absence of competent leaders who understand the complex interaction of these issues, the cooperation needed to ensure a country's future food security may simply not be forthcoming. In the absence of such leadership, a deterioration in the food situation may be unavoidable.

The integration that is needed across the ministries of government is also called for at the international level. Unfortunately, there may be even less contact among the relevant U.N. agencies such as FAO, the U.N. Population Fund (UNFPA), and the U.N. Environment Programme (UNEP) than there is within national ministries. There is no independent water resources agency, nor is there a U.N. agency responsible for transportation. The three U.N. agencies that particularly need to work closely together are FAO, UNFPA, and UNEP.

At another level, the world needs more sophisticated agricultural supply and demand projections. At present, whether they come from FAO, the World Bank, or the USDA, these are largely done by agricultural economists. In a situation where water supplies and temperature levels may have a greater effect on food production in some countries than advancing agricultural technology does, meaningful projections require inputs not only from economists but also from hydrologists, meteorologists, and agronomists.

There is a remarkable lack of data on the status of the world's underground water resources. Few countries systematically gather and report data on changes in water table levels. Even fewer data are available on the thickness of aquifers. And there are almost no projections that tell us when aquifer depletion is likely to occur.

Aside from China, the other big question mark hanging over the world food prospect is Brazil—the most important question being how much of its potential for expanding food production it plans to exploit. Is Brazil prepared to plow the 75 million hectares of the *cerrado* that is believed to be cultivable? Or does it want to preserve part of this land to protect the region's diversity of wildlife and perhaps its rainfall patterns as well? How much of the Amazon is Brazil prepared to clear for agriculture, either for cattle grazing or for crops? What Brazil decides to do in terms of converting the *cerrado* or the Amazon rainforest into cropland and rangeland is directly related to the formulation of population policies in scores of countries. How much should individual countries invest in small-scale water catchment storage, for example? How rigorously should they protect their cropland from conversion to nonfarm uses?[36]

In a world that is increasingly integrated economically, food security is now a global issue. In an integrated world grain market, everyone is affected by the same price shifts. A doubling of grain prices, which is a distinct possibility if we cannot accelerate the growth in grain production, could impoverish more people and destabilize more governments than any event in history. Our future depends on working together to avoid a destabilizing jump in world food prices. Everyone has a stake in stabilizing the agricultural resource base. Everyone has a stake in securing future food supplies. We all have a responsibility to work for the policies—whether in agriculture, energy, population, water use, cropland protection, or soil conservation—that will help ensure future world food security.

The complexity of the challenges the world is facing is matched by the enormity of the effort required to reverse the trends that are undermining future food security. Halting the advancing deserts in China, arresting the fall in water tables in India, and reversing the rise in carbon emissions in the United States are each essential to future world food security. Each will require a strong, new initiative—one that demands a wartime sense of urgency and leadership.

We have inherited the mindset, policies, and fiscal priorities from an era of food security that no longer exists. The policies that once provided food security will no longer suffice in a world where we are pressing against the sustainable yields of oceanic fisheries and underground aquifers and the limits of nature to fix carbon dioxide. Unless we recognize the nature of the era we are entering and adopt new policies and priorities that recognize the earth's natural limits, world food security could begin to deteriorate. If it does, food security could quickly eclipse terrorism as the overriding concern of governments.

Notes

Chapter 1. Pushing Beyond the Earth's Limits

1. United Nations, *World Population Prospects: The 2002 Revision* (New York: 2003).
2. International Monetary Fund (IMF), *World Economic Outlook Database*, at www.imf.org/external/pubs/ft/weo, updated April 2004; Angus Maddison, *Monitoring the World Economy 1820–1992* (Paris: Organisation for Economic Cooperation and Development, 1995).
3. World economy from IMF, op. cit. note 2; water use from Peter H. Gleick, *The World's Water 2000–2001* (Washington, DC: Island Press, 2001), p. 52; demand for seafood from U.N. Food and Agriculture Organization (FAO), *Yearbook of Fishery Statistics* (Rome: various years); carbon emissions from G. Marland, T. A. Boden, and R. J. Andres, "Global, Regional, and National Fossil Fuel CO_2 Emissions," in *Trends: A Compendium of Data on Global Change* (Oak Ridge, TN: Carbon Dioxide Information Analysis Center, Oak Ridge National Laboratory, U.S. Department of Energy, 2003), at cdiac.esd.ornl.gov/trends/trends.htm, updated June 2004.
4. "China Delegation Bought More than 500,000 MT US Wheat–Traders," *Dow Jones Newswires*, 19 February 2004; China National Grain and Oils Information Center, *China Grain Market Weekly Report* (Beijing: 16 April 2004).
5. U.S. Department of Agriculture (USDA), *Production, Supply, and Distribution*, electronic database, at www.fas .usda.gov/psd, updated 13 August 2004.

6. Figure 1–1 compiled from ibid.; wheat prices from IMF, *International Financial Statistics*, electronic database, viewed 2 September 2004; U. S. export restrictions from David Rapp, "Farmer and Uncle Sam: And Old, Odd Couple," *Congressional Quarterly Weekly Report*, 4 April 1987, pp. 598–603.
7. USDA, op. cit. note 5.
8. Ibid.; United Nations, op. cit. note 1.
9. USDA, op. cit. note 5; United Nations, op. cit. note 1.
10. Figure 1–2 compiled from USDA, op. cit. note 5, and from United Nations, op. cit. note 1.
11. L. T. Evans, *Crop Evolution, Adaptation, and Yield* (Cambridge: Cambridge University Press, 1993), pp. 242–44.
12. Fertilizer use from Patrick Heffer, *Short Term Prospects for World Agriculture and Fertilizer Demand 2002/03—2003/04* (Paris: International Fertilizer Industry Association (IFA), 2003); IFA Secretariat and IFA Fertilizer Demand Working Group, *Fertilizer Consumption Report* (Brussels: December 2001); Worldwatch Institute, *Vital Signs 2001* (Washington, DC: 2001); irrigated area from FAO, *FAOSTAT Statistics Database*, at apps.fao.org, updated 2 July 2004.
13. USDA, op. cit. note 5.
14. United Nations, op. cit. note 1.
15. John Krist, "Water Issues Will Dominate California's Agenda This Year," *Environmental News Network*, 21 February 2003.
16. FAO, op. cit. note 3.
17. United Nations, op. cit. note 1.
18. Mark Clayton, "Hunt for Jobs Intensifies as Fishing Industry Implodes," *Christian Science Monitor*, 25 August 1993; Craig S. Smith, "Saudis Worry as They Waste Their Scarce Water," *New York Times*, 26 January 2003; USDA, op. cit. note 5.
19. IMF, op. cit. note 6.
20. USDA, op. cit. note 5; Gleick, op. cit. note 3.
21. Sandra Postel, *Pillar of Sand* (New York: W.W. Norton & Company, 1999); United Nations, op. cit. note 1.
22. World Bank, *China: Agenda for Water Sector Strategy for North China* (Washington, DC: April 2001); Sandra Postel,

Last Oasis (New York: W.W. Norton & Company, 1997), pp. 36–37.

23. Water-to-grain conversion from FAO, *Yield Response to Water* (Rome: 1979); water use from Gleick, op. cit. note 3.

24. John E. Sheehy, International Rice Research Institute, Philippines, e-mail to Janet Larsen, Earth Policy Institute, 2 October 2002.

25. Temperature rise from J. Hansen, NASA's Goddard Institute for Space Studies, "Global Temperature Anomalies in .01 C," at www.giss.nasa.gov/data/update/gistemp; USDA, op. cit. note 5; USDA, *World Agricultural Supply and Demand Estimates* (Washington, DC: 12 August 2003); Janet Larsen, "Record Heat Wave In Europe Takes 35,000 Lives," *Eco-Economy Update* (Washington, DC: Earth Policy Institute, 9 October 2003).

26. Intergovernmental Panel on Climate Change, *Climate Change 2001: The Scientific Basis. Contribution of Working Group I to the Third Assessment Report of the Intergovernmental Panel on Climate Change* (New York: Cambridge University Press, 2001).

27. Grain database at USDA, op. cit. note 5.

28. Figure 1–3 compiled from USDA, op. cit. note 5; Japan multiple cropping index from Ministry of Agriculture, Forestry, and Fisheries, *Statistical Yearbook of Agriculture, Forestry, and Fisheries* (Tokyo: various years).

29. USDA, op. cit. note 5.

30. Ibid.

31. United Nations, op. cit. note 1; IMF, op. cit. note 2; Postel, op. cit. note 21.

32. USDA, Foreign Agricultural Service, *Japan Grain and Feed Annual Report 2003* (Tokyo: March 2003).

33. USDA, op. cit. note 5.

34. Ibid.

35. Ibid.; Susan Cotts Watkins and Jane Menken, "Famines in Historical Perspective," *Population and Development Review*, December 1985.

36. "State Raises Rice Prices Amid Output Drop," *China Daily*, 29 March 2004.

37. USDA, op. cit. note 5.

38. "China Delegation," op. cit. note 4; China National Grain and Oils Information Center, op. cit. note 4.

39. IMF, op. cit. note 2.

40. World Bank, op. cit. note 22, p. vii.

41. Lester R. Brown, "China Losing War With Advancing Deserts," *Eco-Economy Update* (Washington, DC: Earth Policy Institute, 5 August 2003); Wang Tao, Cold and Arid Regions Environmental and Engineering Research Institute, China, e-mail to author, 5 April 2004.

42. "State Raises Rice Prices," op. cit. note 36.

43. USDA, op. cit. note 5.

44. United Nations, op. cit. note 1.

45. World Food Summit from FAO, *The World Food Summit Goal and the Millennium Goals*, Rome, 28 May–1 June 2001, at www.fao.org/docrep/meeting/003/Y0688e.htm; FAO, *The State of Food Insecurity in the World 2003* (Rome: 2003).

46. David B. Lobell and Gregory P. Asner, "Climate and Management Contributions to Recent Trends in U.S. Agricultural Yields," *Science*, 14 February 2003, p. 1032; Shaobing Peng et al., "Rice Yields Decline with Higher Night Temperature from Global Warming," *Proceedings of the National Academy of Sciences*, 6 July 2004, pp. 9971–75.

47. FAO, *The Impact of HIV/AIDS on Food Security, 27th Session of the Committee on World Food Security, Rome*, 28 May–1 June 2001.

48. United Nations, op. cit. note 1; USDA, op. cit. note 5.

49. USDA, op. cit. note 5.

50. United Nations, op. cit. note 1; USDA, op. cit. note 5.

51. USDA, op. cit. note 5.

52. Melissa Alexander, "Focus on Brazil," *World Grain*, January 2004; Marty McVey, "Brazilian Soybeans—Transportation Problems," *AgDM Newsletter*, November 2000.

CHAPTER 2. STOPPING AT SEVEN BILLION

1. United Nations, *World Population Prospects: The 2002 Revision* (New York: 2003).
2. Population Reference Bureau (PRB), *2004 World Population Data Sheet*, wall chart (Washington, DC: 2004).
3. United Nations, op. cit. note 1.
4. R. K. Pachauri and P. V. Sridharan, eds., *Looking Back to Think Ahead (abridged version)*, GREEN India 2047 Project (New Delhi: Tata Energy Research Institute, 1998), p. 7.
5. United Nations, op. cit. note 1.
6. Population projections for Germany, Japan, and Russia from United Nations, op. cit. note 1; Botswana, South Africa, and Swaziland from PRB, op. cit. note 2.
7. Table 2–1 from United Nations, op. cit. note 1.
8. Ibid.
9. Ibid.
10. Ibid.
11. U.S. Department of Agriculture (USDA), *Production, Supply, and Distribution*, electronic database, at www.fas.usda .gov/psd, updated 13 August 2004; United Nations, op. cit. note 1.
12. Figure 2–1 compiled with data from USDA, op. cit. note 11, and from United Nations, op. cit. note 1.
13. Editorial Desk, "Time for Action on Sudan," *New York Times*, 18 June 2004.
14. Ibid.
15. Somini Sengupta, "Where the Land Is a Tinderbox, the Killing Is a Frenzy," *New York Times*, 16 June 2004; Nigeria population data from United Nations, op. cit. note 1; Government of Nigeria, *Combating Desertification and Mitigating the Effects of Drought in Nigeria*, National Report on the Implementation of the United Nations Convention to Combat Desertification (Nigeria: November 1999).
16. Sengupta, op. cit. note 15.
17. Ibid.

18. "Hey! You! Get Off of My Cloud," *Reuters*, 15 July 2004; "Delhi Villagers, Police Clash Over Water Shortage," *The Times of India*, 30 May 2000; "BJP to Launch 'Save Water' Campaign," *The Times of India*, 21 December 1999.

19. United Nations, op. cit. note 1; Sandra Postel, *Pillar of Sand* (New York: W.W. Norton & Company, 1999), pp. 135–36.

20. Frank Notestein, "Population—The Long View," in P. W. Schultz, ed., *Food for the World* (University of Chicago Press: 1945); Figure 2–2 from "How Demographic Transition Reduces Countries' Vulnerability to Civil Conflict," fact sheet (Washington, DC: Population Action International (PAI), updated 17 December 2004).

21. United Nations, op. cit. note 1; USDA, op. cit. note 11.

22. Richard P. Cincotta, Robert Engelman, and Daniele Anastasion, *The Security Demographic: Population and Civil Conflict After the Cold War* (Washington, DC: PAI, 2003), pp. 22–23.

23. Ibid.

24. Ibid., pp. 24–27.

25. Lester R. Brown, Gary Gardner, and Brian Halweil, *Beyond Malthus: Nineteen Dimensions of the Population Challenge* (New York: W.W. Norton & Company, 1999), pp. 111–37.

26. United Nations, op. cit. note 1.

27. Ibid.; U.N. Food and Agriculture Organization Committee on World Food Security, *The Impact of HIV/AIDS on Food Security* (Rome: 2001); Joint United Nations Programme on HIV/AIDS (UNAIDS), *Report on the Global HIV/AIDS Epidemic 2002* (Geneva: July 2002), pp. 44–61.

28. For a discussion of the "demographic bonus," see U.N. Population Fund, *The State of World Population 2002* (New York: 2002), and Cincotta, Engelman, and Anastasion, op. cit. note 22, pp. 33–36.

29. Population growth from United Nations, op. cit. note 1; Japan economic expansion from International Monetary Fund (IMF), *World Economic Outlook Database*, at www.imf.org/external/pubs/ft/weo, updated April 2004.

30. Birth rate data from United Nations, op. cit. note 1; economic expansion from IMF, op. cit. note 29.

31. Birth rate data from United Nations, op. cit. note 1; economic expansion from IMF, op. cit. note 29.
32. Cincotta, Engelman, and Anastasion, op. cit. note 22, p. 36.
33. Ibid., p. 36.
34. Thailand and Iran general country information from *CIA World Factbook*, Online Database, at www.odci.gov/cia/publications/factbook, updated 11 May 2004.
35. G. Tyler Miller, "Cops and Rubbers Day in Thailand," in *Living in the Environment*, 8th ed. (Belmont, CA: Wadsworth Publishing Company, 1994).
36. Ibid.
37. Ibid.; slang for condom from "Mechai Viravaidya: Mr. Condom, Mr. Senator," *Asiaweek*, at www.asiaweek.com/asiaweek/features/power50.2001/p49.html, viewed 10 September 2004.
38. "Mechai Viravaidya: Mr. Condom, Mr. Senator," op. cit. note 37.
39. Miller, op. cit. note 35; demographics from United Nations, op. cit. note 1.
40. Janet Larsen, "Iran's Birth Rate Plummeting at Record Pace," in Lester R. Brown, Janet Larsen, and Bernie Fischlowitz-Roberts, *The Earth Policy Reader* (New York: W.W. Norton & Company, 2002), pp. 190–94; see also Homa Hoodfar and Samad Assadpour, "The Politics of Population Policy in the Islamic Republic of Iran," *Studies in Family Planning*, March 2000, pp. 19–34, and Farzaneh Roudi, "Iran's Family Planning Program: Responding to a Nation's Needs," *MENA Policy Brief*, June 2002; Iran demographics from United Nations, op. cit. note 1.
41. Larsen, op. cit. note 40.
42. Ibid.
43. Ibid.
44. Ibid.
45. United Nations, op. cit. note 1.
46. U.N. General Assembly, "United Nations Millennium Declaration," resolution adopted 18 September 2000; for more

information on the Millennium Development Goals, see www.un.org/millenniumgoals; Paul Blustein, "Global Education Plan Gains Backing," *Washington Post*, 22 April 2002; Gene Sperling, "Educate Them All," *Washington Post*, 20 April 2002; Gene B. Sperling, "Toward Universal Education," *Foreign Affairs*, September/October 2001, pp. 7–13.

47. Susheela Singh et al., *Adding it Up: The Benefits of Investing in Sexual and Reproductive Health Care* (New York: Alan Guttmacher Institute, 2003), pp. 22–25.

48. Jeffrey Sachs, "A New Map of the World," *The Economist*, 22 June 2000; George McGovern, *The Third Freedom: Ending Hunger in Our Time* (New York: Simon & Schuster: 2001), Chapter 1.

Chapter 3. Moving Up the Food Chain Efficiently

1. For data on meat production in various countries, see U.N. Food and Agriculture Organization (FAO), *FAOSTAT Statistics Database*, at apps.fao.org, updated 24 May 2004.

2. Food balance sheets from ibid., updated 27 August 2004.

3. Figure 3–1 compiled from FAO, op. cit. note 1, and from historical statistics in Worldwatch Institute, *Signposts 2002*, CD-Rom (Washington, DC: 2002); average per capita consumption from FAO, op. cit. note 1, and from population in United Nations, *World Population Prospects: The 2002 Revision* (New York: 2003).

4. U.S. Department of Agriculture (USDA), *Production, Supply, and Distribution*, electronic database, at www.fas.usda.gov/psd, updated 13 August 2004; United Nations, op. cit. note 3.

5. USDA, op. cit. note 4; United Nations, op. cit. note 3.

6. USDA, op. cit. note 4; United Nations, op. cit. note 3.

7. FAO, op. cit. note 1; FAO, *FISHSTAT Plus*, electronic database, viewed 13 August 2004.

8. Figure 3–2 compiled from FAO, op. cit. note 1, and from Worldwatch Institute, op. cit. note 3.

9. Dan Murphy, "China Visit Outlines Stark Challenges Ahead For U.S.," at www.meatingplace.com, 13 December 2002.

10. FAO, op. cit. note 1; USDA, op. cit. note 4.

11. FAO, op. cit. note 1.

12. Conversion ratio of grain to beef based on Allen Baker, *Feed Situation and Outlook* staff, Economic Research Service (ERS), USDA, discussion with author, 27 April 1992, on Linda Bailey, *Livestock and Poultry Situation* staff, ERS, USDA, discussion with author, 27 April 1992, and on data taken from various issues of *Feedstuffs*; pork from Leland Southard, *Livestock and Poultry Situation and Outlook* staff, ERS, USDA, discussion with author, 27 April 1992; poultry derived from data in Robert V. Bishop et al., *The World Poultry Market—Government Intervention and Multilateral Policy Reform* (Washington, DC: USDA, 1990); catfish and carp from Rosamond L. Naylor et al., "Effects of Aquaculture on World Fish Supplies," *Nature*, 29 June 2000, p. 1022.

13. Table 3–1 compiled from FAO, op. cit. note 1, and from FAO, op. cit. note 7.

14. FAO, op. cit. note 7; United Nations, op. cit. note 3.

15. Aquaculture output from FAO, op. cit. note 7; S. F. Li, "Aquaculture Research and Its Relation to Development in China," in World Fish Center, *Agricultural Development and the Opportunities for Aquatic Resources Research in China* (Penang, Malaysia: 2001), p. 26; "Mekong Delta to Become Biggest Aquatic Producer in Vietnam," *Vietnam News Agency*, 3 August 2004.

16. Aquaculture from FAO, op. cit. note 7; beef from FAO, op. cit. note 1.

17. USDA, op. cit. note 4; Sadasivam J. Kaushik, "Grain-Based Feeds: The Answer for Aquaculture?" *World Grain*, April 2004.

18. Figure 3–3 compiled from FAO, op. cit. note 1, from FAO, op. cit. note 7, and from Worldwatch Institute, op. cit. note 3; United Nations, op. cit. note 3.

19. FAO, *The State of World Fisheries and Aquaculture 2002* (Rome: 2002), p. 23; Ransom A. Myers and Boris Worm, "Rapid Worldwide Depletion of Predatory Fish Communities," *Nature*, 15 May 2003, pp. 280–83.

20. Myers and Worm, op. cit. note 19; Charles Crosby, "'Blue Frontier' is Decimated," *Dalhousie News*, 11 June 2003.

21. Myers and Worm, op. cit. note 19.
22. Ibid.; Crosby, op. cit. note 20.
23. World Resources Institute, *World Resources 2000–2001* (Washington, DC: 2000).
24. Number of pastoralists from "Investing in Pastoralism," *Agriculture Technology Notes* (Washington, DC: World Bank, Rural Development Department), March 1998, p. 1; FAO, op. cit. note 1.
25. FAO, op. cit. note 1; Worldwatch Institute, op. cit. note 3.
26. USDA, op. cit. note 4; Suzi Fraser Dominy, "Soy's Growing Importance," *World Grain*, 13 April 2004.
27. Author's calculations based on USDA, op. cit. note 4, and USDA, Foreign Agricultural Service (FAS), miscellaneous agricultural reports (Washington, DC: various years).
28. USDA, op. cit. note 4.
29. Ibid.
30. Ibid.; David McKee, "Crushing Competition," *World Grain*, 13 April 2004; USDA, FAS, *China Oilseeds and Products Annual Report 2004* (Beijing: March 2004); Frazer Dominy, op. cit. note 26.
31. USDA, op. cit. note 4.
32. Ibid.; Worldwatch Institute, op. cit. note 3.
33. "Soybean," *Britannica Concise Encyclopedia*, at www.britannica.com/ebc/article?eu=404507, viewed 13 September 2004.
34. USDA, op. cit. note 4.
35. Richard Magleby, "Soil Management and Conservation," in USDA, *Agricultural Resources and Environmental Indicators 2003* (Washington, DC: February 2003), Chapter 4.2, p. 14.
36. Peruvian anchovy industry from Lester R. Brown and Erik P. Eckholm, *By Bread Alone* (New York: Overseas Development Council, 1974), pp. 155–57, and from Marty McVey, Phil Baumel, and Bob Wisner, "Brazilian Soybeans—What is the Potential?" *AgDM Newsletter*, October 2000; expanding soy production and exports from Randall D. Schnepf, Erik N. Dohlman, and Christine Bolling, *Agriculture in Brazil and*

Argentina (Washington, DC: USDA, ERS: 2001), from USDA, op. cit. note 4, and from McVey, Baumel, and Wisner, op. cit. this note.

37. USDA, op. cit. note 4; Worldwatch Institute, op. cit. note 3.

38. USDA, op. cit. note 4.

39. Figure 3–4 compiled from FAO, op. cit. note 1.

40. S. C. Dhall and Meena Dhall, "Dairy Industry—India's Strength in Its Livestock," *Business Line*, Internet Edition of *Financial Daily* from *The Hindu* group of publications, at www.indiaserver.com/businessline/1997/11/07/stories/03070311.htm, 7 November 1997; see also Surinder Sud, "India Is Now World's Largest Milk Producer," *India Perspectives*, May 1999, pp. 25–26; A. Banerjee, "Dairying Systems in India," *World Animal Review*, vol. 79, no. 2 (1994).

41. Milk consumption from FAO, op. cit. note 1; United Nations, op. cit. note 3.

42. Banerjee, op. cit. note 40; Dhall and Dhall, op. cit. note 40.

43. John Wade, Adam Branson, and Xiang Qing, *China Grain and Feed Annual Report 2002* (Beijing: USDA, March 2002); China's crop residue production and use from Gao Tengyun, "Treatment and Utilization of Crop Straw and Stover in China," *Livestock Research for Rural Development*, February 2000.

44. USDA, ERS, "China's Beef Economy: Production, Marketing, Consumption, and Foreign Trade," *International Agriculture and Trade Reports: China* (Washington, DC: July 1998), p. 28.

45. Li, op. cit. note 15, p. 26; FAO, op. cit. note 7.

46. FAO, op. cit. note 1; FAO, op. cit. note 7.

47. United Nations, op. cit. note 3; FAO, op. cit. note 1.

48. China's economic growth from International Monetary Fund, *World Economic Outlook Database*, at www.imf.org/external/pubs/ft/weo, updated April 2004.

49. Figure 3–6 compiled from FAO, op. cit. note 1.

Chapter 4. Raising the Earth's Productivity

1. U.S. Department of Agriculture (USDA), *Production, Supply, and Distribution*, electronic database, at www.fas.usda.gov/psd, updated 13 August 2004; Worldwatch Institute, *Signposts 2002*, CD-Rom (Washington, DC: 2002).
2. Japan, Europe, and U.S. agricultural funding and biotechnology from Margriet F. Caswell et al., *Agricultural Biotechnology: An Economic Perspective* (Washington, DC: USDA, Economic Research Service (ERS), 1998), pp. 15–21; Catalog of Federal Domestic Assistance, "10.203 Payments to Agricultural Experiment Stations Under the Hatch Act," online assistance programs database, at 12.46.245.173/cfda/cfda.html; Aileen Adams et al., *An Assessment of the U.S. Food and Agricultural Research System* (Washington, DC: U.S. Food and Agricultural Research Advisory Panel, 1981), pp. 171–76.
3. USDA, op. cit. note 1.
4. World grain production data from ibid.; 1880s Japan data from "Grain Yields in Japan and India" (USDA, ERS), cited in Lester R. Brown, *Increasing World Food Output: Problems and Prospects*, Foreign Agricultural Economic Report No. 25 (Washington, DC: USDA, ERS, 1965), pp. 13–14.
5. Dwarf wheats and rice information from Thomas R. Sinclair, "Limits to Crop Yield?" in American Society of Agronomy, Crop Science Society of America, and Soil Science Society of America, *Physiology and Determination of Crop Yield* (Madison, WI: 1994), pp. 509–32; India grain production data from USDA, op. cit. note 1.
6. Dwarf wheats and rice information from Sinclair, op. cit. note 5; IRRI data from www.irri.org, viewed 9 September 2004.
7. Percent photosynthate to seed from L. T. Evans, *Crop Evolution, Adaptation and Yield* (Cambridge: Cambridge University Press, 1993), pp. 242–44; theoretical upper limit from Thomas R. Sinclair, "Options for Sustaining and Increasing the Limiting Yield-Plateaus of Grain Crops," paper prepared for the 1998 Symposium on World Food Security, Kyoto, Japan (Washington, DC: USDA Agricultural Research Service, September 1998), p. 14.
8. Evans, op. cit. note 7.

9. Donald N. Duvick, Affiliate Professor of Plant Breeding, Iowa State University, letter to author, 14 March 1997.

10. Evans, op. cit. note 7; Sinclair, op. cit. note 7.

11. Sinclair, op. cit. note 7.

12. USDA, op. cit. note 1; USDA, Foreign Agricultural Service (FAS), *Grains: World Markets and Trade* (Washington, DC: various years).

13. USDA, op. cit. note 1; U.N. Food and Agriculture Organization (FAO), *FAOSTAT Statistics Database*, at apps.fao.org, updated 24 May 2004.

14. USDA, op. cit. note 1; USDA, FAS, *World Agricultural Production* (Washington, DC, various years), at www.fas .usda.gov/wap_arc.html.

15. Table 4–1 from USDA, op. cit. note 1, with France from FAO, op. cit. note 13.

16. USDA, op. cit. note 1.

17. "Justus von Liebig," *Encyclopaedia Britannica* (Cambridge: Encyclopaedia Britannica, Inc., 1976).

18. Figure 4–1 compiled from Patrick Heffer, *Short Term Prospects for World Agriculture and Fertilizer Demand 2002/03 – 2003/04* (Paris: International Fertilizer Industry Association (IFA), December 2003), and from IFA Secretariat and IFA Fertilizer Demand Working Group, *Fertilizer Consumption Report* (Brussels: December 2001), with historical data from Worldwatch Institute, *Signposts 2001*, CD-Rom (Washington, DC: 2001), compiled from IFA and FAO, *Fertilizer Yearbook* (Rome: various years).

19. Figure 4–2 compiled from Heffer, op. cit. note 18, and from IFA Secretariat and IFA Fertilizer Demand Working Group, op. cit. note 18, with historical data from Worldwatch Institute, op. cit. note 18.

20. Heffer, op. cit. note 18; IFA Secretariat and IFA Fertilizer Demand Working Group, op. cit. note 18.

21. Melissa Alexander, "Focus on Brazil," *World Grain*, January 2004.

22. R. J. Diaz, J. Nestlerode, and M. L. Diaz, "A Global Perspective on the Effects of Eutrophication and Hypoxia on

Aquatic Biota," in G. L. Rupp and M. D. White, eds., *Proceedings of the 7th Annual Symposium on Fish Physiology, Toxicology and Water Quality*, Estonia, 12–15 May 2003 (Athens, GA: U.S. Environmental Protection Agency, Ecosystems Research Division: in press).

23. Figure 4–3 compiled from FAO, op. cit. note 13, updated 2 July 2004, and from Worldwatch Institute, op. cit. note 18.

24. FAO, op. cit. note 13, updated 2 July 2004; world grain production data from USDA, op. cit. note 1; China's irrigated land from Worldwatch Institute, op. cit. note 18; Janet Larsen, "Irrigated Area Rises," in Worldwatch Institute, *Vital Signs 2002* (Washington, DC: 2002), pp. 34–35; Lester R. Brown, *Who Will Feed China?* (New York: W.W. Norton & Company, 1995), p. 27.

25. Fertilizer from Heffer, op. cit. note 18 and from IFA Secretariat and IFA Fertilizer Demand Working Group, op. cit. note 18; irrigation from FAO, op. cit. note 13, updated 2 July 2004.

26. USDA, op. cit. note 1; Heffer, op. cit. note 18; IFA Secretariat and IFA Fertilizer Demand Working Group, op. cit. note 18.

27. FAO, op. cit. note 13, updated 2 July 2004.

28. Table 4–2 from USDA, op. cit. note 1; Worldwatch Institute, op. cit. note 1.

29. Figure 4–4 compiled from USDA, op. cit. note 1; monsoon weather from USDA, FAS, *Grains: World Markets and Trade* (Washington, DC: various years).

30. USDA, op. cit. note 1; USDA, op. cit. note 29.

31. USDA, op. cit. note 1.

32. Figure 4–5 compiled from ibid.; France from FAO, op. cit. note 13.

33. Figure 4–6 compiled from USDA, op. cit. note 1; information on China's double cropping in W. Hunter Colby et al., *Agricultural Statistics of the People's Republic of China, 1949-1990* (Washington, DC: USDA, ERS, 1992), and in USDA, FAS, "Crop Calendar," at www.fas.usda.gov/pecad/weather/Crop_calendar/crop_cal.pdf.

34. USDA, op. cit. note 1; U.S. weather from USDA, National Agricultural Statistics Service, "Weekly Weather and Crop

Bulletin," at jan.mannlib.cornell.edu/reports/nassr/field/weather, and from NOAA/ USDA Joint Agricultural Weather Facility, "International Weather and Crop Summary," updated weekly at www.usda.gov/agency/oce/waob/jawf/wwcb/inter.txt.

35. Kenneth G. Cassman et al., "Meeting Cereal Demand While Protecting Natural Resources and Improving Environmental Quality," *Annual Review of Environment and Resources*, November 2003, p. 322.

36. Ibid.

37. Ibid., pp. 324–26.

38. Ibid., pp. 325–26.

39. Ibid., p. 328.

40. Ibid.

41. Ibid., pp. 327–29.

42. Sinclair, op. cit. note 5; Cassman et al., op. cit. note 35.

43. Sinclair, op. cit. note 5; Cassman et al., op. cit. note 35.

44. Sinclair, op. cit. note 5.

45. Kenneth Cassman, Professor and Head of Department of Agronomy and Horticulture, University of Nebraska, letter to author, 7 May 2004.

46. USDA, op. cit. note 1; FAO, op. cit. note 13.

47. Brown, op. cit. note 24, p. 61.

48. Roughage conversion from A. Banerjee, "Dairying Systems in India," *World Animal Review*, vol. 79, no. 2 (1994), and from S. C. Dhall and Meena Dhall, "Dairy Industry—India's Strength in Its Livestock," *Business Line* (Internet Edition of *Financial Daily* from *The Hindu* group of publications), 7 November 1997; China's crop residue production and use from Gao Tengyun, "Treatment and Utilization of Crop Straw and Stover in China," *Livestock Research for Rural Development*, February 2000.

49. Norman E. Borlaug, "The Next Green Revolution," *New York Times*, 11 July 2003.

50. Jules Pretty and Rachel Hine, "Reducing Food Poverty with

Sustainable Agriculture: A Summary of New Evidence," final report from the "SAFE-World" (The Potential of Sustainable Agriculture to Feed the World) Research Project (Colchester, UK: University of Essex, February 2001), p. 11.

51. Ibid.; "Famine in Africa: Controlling Their Own Destiny," (London) *Guardian*, 30 November 2002.

52. Pretty and Hine, op. cit. note 50, p. 21.

Chapter 5. Protecting Cropland

1. Ann Schrader, "Latest Import From China: Haze," *Denver Post*, 18 April 2001; Paul Brown, "4x4s Replace the Desert Camel and Whip Up a Worldwide Dust Storm," (London) *Guardian*, 20 August 2004.

2. Howard W. French, "China's Growing Deserts Are Suffocating Korea," *New York Times*, 14 April 2002.

3. Brown, op. cit. note 1.

4. Calculations for paved area by Janet Larsen, Earth Policy Institute, based on U.S. Department of Transportation, Federal Highway Administration (FHWA), *Highway Statistics 1999* (Washington, DC: 2001), on Mark Delucchi, "Motor Vehicle Infrastructure and Services Provided by the Public Sector," cited in Todd Litman, *Transportation Land Valuation* (Victoria, BC, Canada: Victoria Transport Policy Institute, 2000), p. 4, on Ward's Communications, *Ward's World Motor Vehicle Data* (Southfield, MI: 2000), on Jeffrey Kenworthy, Associate Professor in Sustainable Settlements, Institute for Sustainability and Technology Policy, Murdoch University, Australia, e-mail message to Larsen, and on David Walterscheid, FHWA Real Estate Office, discussion with Larsen.

5. United Nations, *World Population Prospects: The 2002 Revision* (New York: 2003); land requirements are author's estimate.

6. United Nations, op. cit. note 5.

7. Livestock data from U.N. Food and Agriculture Organization (FAO), *FAOSTAT Statistics Database*, at apps.fao.org, updated 24 May 2004.

8. Yang Youlin, Victor Squires, and Lu Qi, eds., *Global Alarm:*

Dust and Sandstorms from the World's Drylands (Bangkok: Secretariat of the U.N. Convention to Combat Desertification, September 2002), p. 319.

9. U.S. Department of Agriculture (USDA), *Summary Report: 1997 Natural Resources Inventory* (Washington, DC: 1999, rev. 2000), pp. 46–51.
10. Table 5–1 from U.N. Environment Programme (UNEP), cited in Global Environment Fund–International Fund for Agricultural Development Partnership, *Tackling Land Degradation and Desertification* (Washington, DC: July 2002); grainland from USDA, *Production, Supply, and Distribution,* electronic database, at www.fas.usda.gov/psd, updated 13 August 2004; population from United Nations, op. cit. note 5.
11. John Steinbeck, *The Grapes of Wrath* (New York: Penguin Books, 1992).
12. FAO, *The State of Food and Agriculture 1995* (Rome: 1995), p. 175.
13. USDA, op. cit. note 10.
14. U.S. Embassy, *Grapes of Wrath in Inner Mongolia* (Beijing: May 2001).
15. Brown, op. cit. note 1.
16. Ibid.
17. UNEP, *Afghanistan: Post-Conflict Environmental Assessment* (Geneva: 2003), pp. 50–60; NASA, "Dust Storm Over Southern Asia," Earth Observatory Newsroom, at earth observatory.nasa.gov/Newsroom, 6 May 2004.
18. World Resources Institute, *World Resources 2000–2001* (Washington, DC: 2000). Table 5–2 from the following: UNEP, op. cit. note 17; Government of Brazil, *National Report on the Implementation of the United Nations Convention to Combat Desertification* (Brasilia: 2002); Expert Group for Compiling National Report for Implementing the UNCCD, *China National Report to Implement the United Nations Convention to Combat Desertification* (Beijing: 16 April 2002); Ministry of Environment and Forests of India, *Second National Report on Implementation of the United Nations Convention to Combat Desertification* (New Delhi: 30 April 2002); Iranian News Agency, "Official Warns of

Impending Desertification Catastrophe in Southeast Iran," *BBC International Reports*, 29 September 2002; Republic of Kenya, Ministry of Environment and Natural Resources, *National Action Programme: A Framework for Combating Desertification in Kenya in the Context of the United Nations Convention to Combat Desertification* (Nairobi: February 2002); Mexico from "Desertification is Both a Cause and a Consequence of Poverty," *Environmental News Service*, 17 June 2003; Government of Nigeria, *Combating Desertification and Mitigating the Effects of Drought in Nigeria,* National Report on the Implementation of the United Nations Convention to Combat Desertification (Nigeria: November 1999); Government of Yemen, *National Report on the Implementation of the United Nations Convention to Combat Desertification* (Sana'a, Yemen: 2000).

19. Government of Nigeria, op. cit. note 18; population from United Nations, op. cit. note 5; livestock from FAO, op. cit. note 7.

20. Government of Nigeria, op. cit. note 18; population from United Nations, op. cit. note 5.

21. "Case Studies of Sand-Dust Storms in Africa and Australia," in Yang, Squires, and Lu, eds., op. cit. note 8, pp. 123–66.

22. "Algeria to Convert Large Cereal Land to Tree-Planting," *Reuters*, 8 December 2000; imports from USDA, op. cit. note 10.

23. Economic losses from desertification in China from "Summaries of Reports Submitted by Selected Asian Country Parties," United Nations Convention to Combat Desertification, Conference of the Parties, Fourth Session, 11–22 December 2000 (Bonn: 12 December 2000), p. 18; "Desert Mergers and Acquisitions," *Beijing Environment, Science, and Technology Update*, U.S. Embassy in China, 19 July 2002, p. 2.

24. "Desert Mergers and Acquisitions," op. cit. note 23; Zhang Tingting, "Xinjiang Deserts Moving Closer," *China Internet Information Center*, 27 June 2002; photographs from Lu Tongjing, *Desert Witness: Images of Environmental Degradation in China's Northwest* (Beijing: Heinrich Boll Foundation and China Environment and Sustainable Development Reference and Research Center).

25. United Nations, op. cit. note 5; land requirements are author's estimate.

26. Carl J. Dahlman, *China and the Knowledge Economy: Seizing the 21st Century* (Washington, DC: World Bank, January 2002); United Nations, op. cit. note 5.

27. United Nations, op. cit. note 5.

28. Figure 5–1 from Worldwatch Institute, *Signposts 2004, CD-Rom* (Washington, DC: 2004); calculations for paved area by Larsen, op. cit. note 4.

29. Vehicle fleet from Ward's Communications, op. cit. note 4; population from United Nations, op. cit. note 5; grain from USDA, op. cit. note 10.

30. Calculations for paved area by Larsen, op. cit. note 4.

31. Ibid.

32. Worldwatch Institute, op. cit. note 28; Ward's Communications, op. cit. note 4.

33. U. S. experience in USDA, Economic Research Service, *Agri-Environmental Policy at the Crossroads: Guideposts on a Changing Landscape*, Agricultural Economic Report No. 794 (Washington, DC: January 2001), p. 16; loss of topsoil from water erosion from USDA, op. cit. note 9; China from Chen Xiwen, Deputy Director, Development Research Center of the State Council, and colleagues, discussion with author in Beijing, 16 May 2002.

34. R. Neil Sampson, *Farmland or Wasteland* (Emmaus, PA: Rodale Press, 1981), p. 242.

35. USDA, Natural Resources Conservation Service, *CORE4 Conservation Practices Training Guide: The Common Sense Approach to Natural Resource Conservation* (Washington, DC: August 1999); Rolf Derpsch, "Frontiers in Conservation Tillage and Advances in Conservation Practice," in D. E. Stott, R. H. Mohtar, and G. C. Steinhardt, eds., *Sustaining the Global Farm*, selected papers from the 10th International Soil Conservation Organization Meeting, at Purdue University and USDA-ARS National Soil Erosion Research Laboratory, 24–29 May 1999 (Washington, DC: 2001), pp. 248–54.

36. Table 5–3 from Rolf Derpsch and J. R. Benites, "Agricultura Conservacionista no Mundo," presented at the Brazilian Soil

Science Conference, Santa Maria, Brazil, July 2004.

37. USDA, Farm Service Agency Online, "History of the CRP," in *The Conservation Reserve Program*, at www.fsa.usda.gov/dafp/cepd/12crplogo/history.htm, viewed 16 September, 2004; Derpsch and Benites, op. cit. note 36; Conservation Technology Information Center (CTIC), Purdue University, "1990–2002 Conservation Tillage Trends," from 2002 National Crop Residue Management Survey, at www.ctic.purdue.edu/Core4/CT/CTSurvey/NationalData.html, viewed 23 September 2004; CTIC, Purdue University, "Crop Residue Management," National Crop Residue Management Survey Data, at www.ctic.purdue.edu/Core4/CT/CT.html, viewed 23 September 2004.

38. CTIC, "1990–2002 Conservation Tillage Trends," op. cit. note 37; Derpsch, op. cit. note 35; FAO, "Intensifying Crop Production With Conservation Agriculture," at www.fao.org/waicent/faoinfo/agricult/ags/AGSE/agse_e/general/CONT1.htm, viewed 30 September 2004.

39. John Whitlegg, editorial, *World Transport Policy and Practice*, vol. 8, no. 4 (2002), p. 5; Randy Kennedy, "The Day the Traffic Disappeared," *New York Times Magazine*, 20 April 2003, pp. 42–45.

40. Ding Guangwei and Li Shishun, "Analysis of Impetuses to Change of Agricultural Land Uses in China," *Bulletin of the Chinese Academy of Sciences*, vol. 13, no. 1 (1999).

41. Ibid.

42. Anthonie Gerard Welleman, project manager of the Bicycle Master Plan at the Dutch Ministry of Transport, Public Works and Water Management, presentation at the Velo-City Conference '95 (Basel, Switzerland: 1995), at www.communitybike.org/cache/autumn_bike_master_plan.html.

43. United Nations, op. cit. note 5.

Chapter 6. Stabilizing Water Tables

1. Jacob W. Kijne, *Unlocking the Water Potential of Agriculture* (Rome: U.N. Food and Agriculture Organization (FAO), 2003), p. 26; calculation based on 1,000 tons of water for 1 ton of grain from FAO, *Yield Response to Water* (Rome: 1979).

2. Water use from Peter H. Gleick, *The World's Water 2000–2001* (Washington, DC: Island Press, 2000), p. 52.

3. Ibid.

4. Grain production from U.S. Department of Agriculture (USDA), *Production, Supply, and Distribution*, electronic database, at www.fas.usda.gov/psd, updated 13 August 2004; population from United Nations, *World Population Prospects: The 2002 Revision* (New York: 2003). Table 6–1 from the following: Mexico from Yacov Tsur et al., *Pricing Irrigation Water: Principles and Cases from Developing Countries* (Washington, DC: Resources For the Future, 2004), pp. 90–91, 218, 222; U.S. aquifer information from "High Plains Aquifer Down by Six Percent," *Environment New Service*, 11 February 2004, from U.S. Geological Survey (USGS), "High Plains Regional Groundwater Study," National Water Quality Assessment Program, 15 January 2004, from Carey Gillam, "Fight on to Save Plains Water Source," *Reuters*, 1 December 2003, and from USDA, *Agricultural Resources and Environmental Indicators 2000* (Washington, DC: 2000), Chapter 2.1, pp. 5–6; Saudi Arabia wheat production data from USDA, *Production, Supply, and Distribution*, op. cit. this note; Saudi aquifer information from Craig S. Smith, "Saudis Worry as They Waste Their Scarce Water," *New York Times*, 26 January 2003; Iran overpumping from Chenaran Agricultural Center, Ministry of Agriculture, according to Hamid Taravati, publisher, Iran, e-mail to author, 25 June 2002; Yemen's water situation from Christopher Ward, "Yemen's Water Crisis," based on a lecture to the British Yemeni Society in September 2000, July 2001, from Christopher Ward, *The Political Economy of Irrigation Water Pricing in Yemen* (Sana'a, Yemen: World Bank, November 1998), and from Marcus Moench, "Groundwater: Potential and Constraints," in Ruth S. Meinzen-Dick and Mark W. Rosegrant, eds., *Overcoming Water Scarcity and Quality Constraints* (Washington, DC: International Food Policy Research Institute (IFPRI), October 2001); Israel water tables from Deborah Camiel, "Israel, Palestinian Water Resources Down the Drain," *Reuters*, 12 July 2000; India water depletion from Tushaar Shah et al., *The Global Groundwater Situation: Overview of Opportunities and Challenges* (Colombo, Sri Lanka: International Water Management Institute (IWMI), 2000), and from David Seckler, David Molden, and Randolph Barker, "Water

Scarcity in the Twenty-First Century," *Water Brief 1* (Colombo, Sri Lanka: IWMI, 1999), p. 2; North China Plain from Michael Ma, "Northern Cities Sinking as Water Table Falls," *South China Morning Post*, 11 August 2001, and from Hong Yang and Alexander Zehnder, "China's Regional Water Scarcity and Implications for Grain Supply and Trade," *Environment and Planning A*, vol. 33 (2001); China's wheat imports from USDA, *Production, Supply, and Distribution*, op. cit. this note.

5. U.S. water depletion from USDA, *Agricultural Resources and Environmental Indicators 2000*, op. cit. note 4, p. 6; India water depletion from Shah et al., op. cit. note 4, and from Seckler, Molden, and Barker, op. cit. note 4; North China Plain from Ma, op. cit. note 4; share of China's grain harvest from the North China Plain based on Yang and Zehnder, op. cit. note 4, and on USDA, *Production, Supply, and Distribution*, op. cit. note 4.

6. Ogallala aquifer information from "High Plains Aquifer Down by Six Percent," op. cit. note 4, from USGS, op. cit. note 4, from Gillam, op. cit. note 4, and from USDA, *Agricultural Resources and Environmental Indicators 2000*, op. cit. note 4; Saudi Arabia aquifer information from Smith, op. cit. note 4; deep North China Plain aquifer information from Ma, op. cit. note 4.

7. Figure 6–1 compiled from USDA, *Production, Supply, and Distribution*, op. cit. note 4.

8. Ibid.

9. Fred Pearce, "Asian Farmers Sucking the Continent Dry," *New Scientist*, 25 August 2004.

10. Ibid.; Tamil Nadu population from 2001 census, "Tamil Nadu at a Glance: Area and Population" at www.tn.gov.in.

11. Pearce, op. cit. note 9.

12. Yacov Tsur et al., *Pricing Irrigation Water: Principles and Cases from Developing Countries* (Washington, DC: Resources for the Future, 2004), p. 219.

13. "High Plains Aquifer Down by Six Percent," op. cit. note 4; USGS, op. cit. note 4; Gillam, op. cit. note 4; USDA, *Agricultural Resources and Environmental Indicators 2000*, op. cit.

note 4; Sandra Postel, *Pillar of Sand* (New York: W.W. Norton & Company, 1999), p. 77.

14. Table 6–2 from the following: Amu Darya, Colorado, Ganges, Indus, and Nile rivers from Postel, op. cit. note 13, pp. 59, 71–73, 94, 261–62; Fen and Yellow Rivers from Lester R. Brown and Brian Halweil, "China's Water Shortages Could Shake World Food Security," *World Watch*, July/August 1998, p. 11; India Ganges population from Carmen Revenga et al., *Watersheds of the World* (Washington, DC: World Resources Institute and Worldwatch Institute, 1998), Section 2-78; population projections from United Nations, op. cit. note 4.

15. Fen River/Taiyuan from Brown and Halweil, op. cit. note 14.

16. Postel, op. cit. note 13, pp. 141–49.

17. Population projections from United Nations, op. cit. note 4.

18. Ethiopia and Egypt incomes from International Monetary Fund (IMF), *World Economic Outlook Database*, at www.imf.org/external/pubs/ft/weo, updated April 2004; population from United Nations, op. cit. note 4; Ethiopia water development plans from Postel, op. cit. note 13, pp. 143–44.

19. Moench, op. cit. note 4.

20. Postel, op. cit. note 13, pp. 59, 94.

21. U.N. Environment Programme (UNEP), "'Garden of Eden' in Southern Iraq Likely to Disappear Completely in Five Years Unless Urgent Action Taken," news release (Nairobi: 22 March 2003); Hassan Partow, *The Mesopotamian Marshlands: Demise of an Ecosystem*, Early Warning and Assessment Technical Report (Nairobi: Division of Early Warning and Assessment, UNEP, 2001); population from United Nations, op. cit. note 4.

22. Noel Gollehon and William Quinby, "Irrigation in the American West: Area, Water and Economic Activity," *Water Resources Development*, vol. 16, no. 2 (2000), pp. 187–95; Patrick O'Driscoll, "Dry West Sends Out for Water," *USA Today*, 27 July 2004.

23. Joey Bunch, "Cities' Water Needs Uprooting Colorado Farms," *Denver Post*, 11 July 2004; *The Water Strategist*, various issues at www.waterstrategist.com.

24. Arkansas River basin from Joey Bunch, "Water Projects Fore-

cast to Fall Short of Needs: Study Predicts 10% Deficit in State," *Denver Post*, 22 July 2004.

25. Dean Murphy, "Pact in West Will Send Farms' Water to Cities," *New York Times*, 17 October 2003; Tim Molloy, "California Water District Approves Plan to Pay Farmers for Irrigation Water," *Associated Press*, 13 May 2004; California population from Californians for Population Stabilization, population figures, at www.cap-s.org, viewed 1 October 2004.

26. "China Politics: Growing Tensions Over Scarce Water," *The Economist*, 21 June 2004.

27. Calculation based on 1,000 tons of water for 1 ton of grain from FAO, op. cit. note 1, and on global wheat prices from IMF, *International Financial Statistics* (Washington, DC: various years); industrial water intensity in Mark W. Rosegrant, Claudia Ringler, and Roberta V. Gerpacio, "Water and Land Resources and Global Food Supply," paper prepared for the 23rd International Conference of Agricultural Economists on Food Security, Diversification, and Resource Management: Refocusing the Role of Agriculture, Sacramento, CA, 10–16 August 1997.

28. Water-to-grain conversion from FAO, op. cit. note 1.

29. Hans Lofgren and Alan Richards, "Food Security, Poverty, and Economic Policy in the Middle East and North Africa," *TMD Discussion Paper 111* (Washington, DC: International Food Policy Research Institute, February 2003); IWMI, "Trade as a Means to Food and Water Security," *GWP Issues Paper* (Sri Lanka: 13–16 December 2000).

30. Nile River flow from Postel, op. cit. note 13; grain imports from USDA, *Production, Supply, and Distribution*, op. cit. note 4; calculation based on 1,000 tons of water for 1 ton of grain from FAO, op. cit. note 1.

31. Grainland productivity from USDA, *Production, Supply, and Distribution*, op. cit. note 4; Worldwatch Institute, *Signposts 2002*, CD-Rom (Washington, DC: 2002).

32. Calculation based on 1,000 tons of water for 1 ton of grain from FAO, op. cit. note 1.

33. Sandra Postel and Amy Vickers, "Boosting Water Productivity," in Worldwatch Institute, *State of the World 2004*

(New York: W.W. Norton & Company, 2004), pp. 51–52; Gleick, op. cit. note 2.

34. Wang Shucheng, private meeting with author, Beijing, May 2004.

35. FAO, *Crops and Drops* (Rome: 2002), p. 17; Alain Vidal, Aline Comeau, and Hervé Plusquellec, *Case Studies on Water Conservation in the Mediterranean Region* (Rome: FAO, 2001), p. vii.

36. FAO, op. cit. note 35; Vidal, Comeau, and Plusquellec, op. cit. note 35.

37. Table 6–3 from Postel and Vickers, op. cit. note 33, p. 53.

38. Peter Wonacott, "To Save Water, China Lifts Price," *Wall Street Journal*, 14 June 2004.

39. Greg Leslie, "Solving Water Problem Could Come Down To Using City's Kidneys," *Sydney Morning Herald*, 19 April 2004.

40. USGS, "Estimated Water Use in the United States," news briefing (Reston, VA: March 2004).

41. USDA, *Production, Supply, and Distribution*, op. cit. note 4.

42. Population from United Nations, op. cit. note 4; grain consumption from USDA, *Production, Supply, and Distribution*, op. cit. note 4; water calculation based on 1,000 tons of water for 1 ton of grain from FAO, op. cit. note 1.

43. United Nations, op. cit. note 4; for countries overpumping aquifers see note 4, this chapter, and Lester R. Brown, *Plan B: Rescuing a Planet under Stress and a Civilization in Trouble* (New York: W.W. Norton & Company, 2003), p. 26.

Chapter 7. Stabilizing Climate

1. Shaobing Peng et al., "Rice Yields Decline With Higher Night Temperature From Global Warming," *Proceedings of the National Academy of Sciences*, 6 July 2004, pp. 9971–75.

2. U.S. Department of Agriculture (USDA), *Production, Supply, and Distribution*, electronic database, at www.fas.usda .gov/psd, updated 13 August 2004; Janet Larsen, "Record Heat Wave In Europe Takes 35,000 Lives," *Eco-Economy*

Update (Earth Policy Institute), 9 October 2003.

3. United Nations, *World Population Prospects: The 2002 Revision* (New York: 2003).

4. Peng et al., op. cit. note 1.

5. Ibid.

6. David B. Lobell and Gregory P. Asner, "Climate and Management Contributions to Recent Trends in U.S. Agricultural Yields," *Science*, 14 February 2003, p. 1032.

7. John E. Sheehy, International Rice Research Institute, Philippines, e-mail to Janet Larsen, Earth Policy Institute, 2 October 2002.

8. L. H. Allen, Jr., et al., "Carbon Dioxide and Temperature Effects on Rice," in S. Peng et al., eds., *Climate Change and Rice* (Berlin: Springer-Verlag, 1995), pp. 258–77; B. A. Kimball, K. Kobayashi, and M. Bindi, "Responses of Agricultural Crops to Free-Air CO_2 Enrichment," *Advances in Agronomy*, vol. 77, pp. 293–368; K. S. Kavi Kumar and Jyoti Parikh, "Socio-Economic Impacts of Climate Change on Indian Agriculture," *International Review for Environmental Strategies*, vol. 2, no. 2 (2001), pp. 277–93.

9. Mohan K. Wali et al., "Assessing Terrestrial Ecosystem Sustainability," *Nature & Resources*, October–December 1999, pp. 21–33.

10. USDA, *World Agricultural Supply and Demand Estimates* (Washington, DC: 12 August 2003), p. 1.

11. Peter Griffiths, "Record Heatwave Bakes Britain," *Reuters*, 12 August 2003; Alessandra Rizzo, "Italy Heat Wave Said to Kill 4,000 Elders," *Associated Press*, 11 September 2003; Daniel Silva, "Portugal Heat-Wave Toll Well Below Initial Estimate of 1,300: Minister," *Agence France-Presse*, 3 September 2003; National Atmospheric and Oceanic Administration/USDA, *International Weather and Crop Summary* (Washington, DC: 29 June 2003); World Meteorological Organization, "Global Temperature in 2003 Third Warmest," press release, 22 December 2003.

12. Eastern Europe wheat crop from USDA, *Grain: World Markets and Trade* (Washington, DC: August 2003), p. 8; Ukraine and Romania wheat crop and Ukraine forced to import from USDA, op. cit. note 2; Czech Republic grain harvest from

Mark Baker, "Europe: Crops Withering in Searing Heat," *Radio Free Europe/Radio Liberty*, 14 August 2003.

13. Larsen, op. cit. note 2.

14. Figure 7–1 from J. Hansen, NASA's Goddard Institute for Space Studies (GISS), "Global Temperature Anomalies in .01 C," at www.giss.nasa.gov/data/update/gistemp, viewed 16 September 2004; USDA, op. cit. note 2.

15. Figure 7–2 from C. D. Keeling, T. P. Whorf, and the Carbon Dioxide Research Group, "Atmospheric Carbon Dioxide Record from Mauna Loa," Scripps Institution of Oceanography, University of California, at cdiac.esd.ornl.gov/trends/trends.htm, updated June 2004; Intergovernmental Panel on Climate Change (IPCC), *Climate Change 2001: The Scientific Basis. Contribution of Working Group I to the Third Assessment Report* (New York: Cambridge University Press, 2001).

16. IPCC, op. cit. note 15.

17. John Krist, "Water Issues Will Dominate California's Agenda This Year," *Environmental News Network*, 21 February 2003.

18. World Wide Fund for Nature, "Going, Going, Gone! Climate Change and Global Glacier Decline," news release, at www.panda.org/about_wwf/what_we_do/climate_change/problems/impacts_glaciers.cfm, 27 November 2003; Global Land Ice Measurements from Space, "Decline of World's Glaciers Expected to Have Global Impacts Over this Century," NASA Goddard Space Flight Center, news release, 29 May 2002.

19. IPCC, *Climate Change 2001: Impacts, Adaptation, and Vulnerability. Contribution of Working Group II to the Third Assessment Report* (New York: Cambridge University Press, 2001).

20. Ibid.

21. IPCC, op. cit. note 15; University of Colorado at Boulder, "Global Sea Levels Likely to Rise in 21st Century Than Previous Predictions," press release (Boulder, CO: 16 February 2002).

22. W. Krabill et al., "Greenland Ice Sheet: High Elevation Balance and Peripheral Thinning," *Science*, 21 July 2000, p. 428.

23. World Bank, *World Development Report 1999/2000* (New

York: Oxford University Press, 2000), p. 100; population from United Nations, op. cit. note 3.

24. IPCC, op. cit. note 15.

25. USDA, op. cit. note 2.

26. U.S. Department of Energy and Environmental Protection Agency, *Fuel Economy Guide* (Washington, DC: 2004); gasoline savings based on Malcolm A. Weiss et al., *Comparative Assessment of Fuel Cell Cars* (Cambridge, MA: Massachusetts Institute of Technology, 2003).

27. U.S. Department of Transportation, Bureau of Transportation Statistics, *Transportation Statistics Annual Report* (Washington, DC: October 2003), p. 52.

28. D. L. Elliott, L. L. Wendell, and G. L. Gower, *An Assessment of the Available Windy Land Area and Wind Energy Potential in the Contiguous United States* (Richland, WA: Pacific Northwest Laboratory, 1991); C. L. Archer, and M. Z. Jacobson, "The Spatial and Temporal Distributions of U.S. Winds and Wind Power at 80 m Derived From Measurements," *Journal of Geophysical Research*, 16 May 2003.

29. Howard Geller, *Compact Fluorescent Lighting*, American Council for an Energy Efficient Economy Technology Brief, at www.aceee.org, viewed 17 September 2004.

30. Elliott, Wendell, and Gower, op. cit. note 28.

31. Ibid.; Archer and Jacobson, op. cit. note 28.

32. European Wind Energy Association (EWEA), *Wind Power Targets for Europe: 75,000 MW by 2010* (Belgium: 2003).

33. Garrad Hassan and Partners, Ltd., *Sea Wind Europe* (London: Greenpeace, 2004).

34. Figure 7–3 from Worldwatch Institute, *Signposts 2001*, CD-Rom (Washington, DC: 2001), updated by Earth Policy Institute from American Wind Energy Association (AWEA), *Global Wind Energy Market Report* (Washington, DC: updated March 2004); oil from Energy Information Administration, "World Oil Demand," *International Petroleum Monthly*, April 2004; natural gas and coal from Janet L. Sawin, "Fossil Fuel Use Up," in Worldwatch Institute, *Vital Signs 2003* (New York: W.W. Norton & Company, 2003), pp. 34–35; nuclear power from Nicholas Lenssen, "Nuclear

Power Rises," in Worldwatch Institute, *Vital Signs 2003*, op. cit. this note.

35. Worldwatch Institute, *Signposts 2001*, op. cit. note 34, updated by Earth Policy Institute from AWEA, op. cit. note 34; EWEA, *Europe's Installed Wind Capacity* (Brussels: 2003); Denmark from Soren Krohn, "Wind Energy Policy in Denmark: Status 2002," Danish Wind Energy Association, at www.windpower.org/articles/energypo.htm, February 2002.

36. Larry Flowers, National Renewable Energy Laboratory, "Wind Power Update," at www.eren.doe.gov/windpowering america/pdfs/wpa/wpa_update.pdf, viewed 19 June 2002; Glenn Hasek, "Powering the Future," *Industry Week*, 1 May 2000; 2¢ per kilowatt-hour from EWEA and Greenpeace, *Wind Force 12* (Brussels: 2003).

37. "US Wind Power Industry Gets Tax Credit Boost," *Reuters*, 13 March 2002; "Blocked US Energy Bill Slows Wind Power Projects," *Reuters*, 12 January 2004.

Chapter 8. Reversing China's Harvest Decline

1. U.S. Department of Agriculture (USDA), *Production, Supply, and Distribution,* electronic database, at www.fas.usda .gov/psd, updated 13 August 2004; historical data from Worldwatch Institute, *Signposts 2001*, CD-Rom (Washington, DC: 2001).

2. USDA, op. cit. note 1; Wang Tao, "The Process and Its Control of Sandy Desertification in Northern China," seminar on desertification in China, Cold and Arid Regions Environmental & Engineering Institute, Chinese Academy of Sciences, Lanzhou, China, May 2002.

3. USDA, op. cit. note 1; "State Raises Rice Prices Amid Output Drop," *China Daily*, 29 March 2004.

4. USDA, op. cit. note 1.

5. Population from United Nations, *World Population Prospects: The 2002 Revision* (New York: 2003); livestock population from U.N. Food and Agriculture Organization (FAO), *FAOSTAT Statistics Database*, at apps.fao.org, updated 24 May 2004; desertification in China from Wang, op. cit. note 2; U.S. Embassy, "Desert Mergers and Acquisitions,"

Beijing Environment, Science, and Technology Update (Beijing: 19 July 2002).

6. Wang, op. cit. note 2.
7. Wang Tao, Cold and Arid Regions Environmental & Engineering Institute, e-mail message to author, 6 April 2004.
8. Table 8–1 from China Meteorological Administration, cited in U.S. Embassy, *Grapes of Wrath in Inner Mongolia* (Beijing: 2001).
9. FAO, op. cit. note 5; historical data from Worldwatch Institute, op. cit. note 1; livestock concentration from Hu Zizhi and Zhang Degang, "China," *Country Pasture/Forage Resource Profiles*, at www.fao.org/WAICENT/FAOINFO/AGRICULT/AGP/AGPC/doc/Counprof/china/china1.htm, viewed 30 September 2004.
10. FAO, op. cit. note 5; historical data from Worldwatch Institute, op. cit. note 1; photograph from Lu Tongjing, *Desert Witness: Images of Environmental Degradation in China's Northwest* (Beijing: Heinrich Boll Foundation and China Environment and Sustainable Development Reference and Research Center, 2003).
11. Asian Development Bank, *Technical Assistance to The People's Republic of China for Optimizing Initiatives to Combat Desertification in Gansu Province* (Manila, Philippines: 2001).
12. U.S. Embassy, op. cit. note 8.
13. U.S. Bureau of the Census, *1930 Fact Sheet*, at www.census.gov, revised March 2002; United Nations, op. cit. note 5.
14. China Internet Information Center, "National Conditions" at www.china.org.cn/english/shuzi-en/en-shuzi/gq/htm/s .htm, viewed 1 September 2004.
15. "China Publishes Annual Report on Land and Resources," *Xinhua News Agency*, 9 April 2004.
16. Calculations for paved area by Janet Larsen, Earth Policy Institute, based on U.S. Department of Transportation, Federal Highway Administration (FHWA), *Highway Statistics 1999* (Washington, DC: 2001), on Mark Delucchi, "Motor Vehicle Infrastructure and Services Provided by the Public Sector," cited in Todd Litman, *Transportation Land Valuation*

(Victoria, B.C., Canada: Victoria Transport Policy Institute, November 2000), p. 4, on Ward's Communications, *Ward's World Motor Vehicle Data* (Southfield, MI: 2000), on Jeffrey Kenworthy, Associate Professor in Sustainable Settlements, Institute for Sustainability and Technology Policy, Murdoch University, Australia, e-mail message to Larsen, and on David Walterscheid, FHWA Real Estate Office, discussion with Larsen; cars sold in China from Peter S. Goodman, "Car Culture Captivates China," *Washington Post*, 8 March 2004; estimated grain production from USDA, op. cit. note 1.

17. Calculations for paved area by Larsen, op. cit. note 16; grainland from USDA, op. cit. note 1.
18. Figure 8–2 from FAO, op. cit. note 5; average farm size from Roy L. Prosterman, "China's New Market in Land," *Wall Street Journal*, 7 March 2003.
19. Ministry of Agriculture, Forestry, and Fisheries, *Statistical Yearbook of Agriculture, Forestry, and Fisheries* (Tokyo: various years); Ministry of Agriculture, Forestry, and Fisheries, *Statistical Yearbook of Agriculture, Forestry, and Fisheries* (Seoul: various years); Taiwan from John Dyck, USDA, Foreign Agricultural Service, Washington, DC, private communication, 16 March 1995.
20. "State Raises Rice Prices," op. cit. note 3; "Unprecedented State Subsidy Spurs China's Grain Production," *World News Connection*, 17 April 2004.
21. FAO, *FISHSTAT Plus*, electronic database, viewed 13 September 2004; Jinyun Ye, *Carp Polyculture System in China* (Huzhou, China: Institute of Freshwater Fisheries, 1998).
22. FAO, op. cit. note 21.
23. Casey E. Bean and Adam Branson, *Fishery Products Situation Report* (Beijing: USDA, Foreign Agricultural Service, 17 June 2004); Suzi Fraser Dominy, "Soy's Growing Importance," *World Grain*, 1 June 2003.
24. FAO, op. cit. note 21.
25. USDA, Foreign Agricultural Service, *China Oilseeds and Products Annual Report 2004* (Beijing: March 2004); Fraser Dominy, op. cit. note 23; USDA, op. cit. note 1.
26. Bean and Branson, op. cit. note 23.

27. Ibid.

28. World Bank, *China: Agenda for Water Sector Strategy for North China* (Washington, DC: 2001); Sandra Postel, *Last Oasis* (New York: W.W. Norton & Company, 1997), p. 36–37; share of China's grain harvest from the North China Plain based on Hong Yang and Alexander Zehnder, "China's Regional Water Scarcity and Implications for Grain Supply and Trade," *Environment and Planning A*, vol. 33 (2001), and on USDA, op. cit. note 1.

29. Michael Ma, "Northern Cities Sinking as Water Table Falls," *South China Morning Post*, 11 August 2001.

30. Lester R. Brown and Brian Halweil, "China's Water Shortages Could Shake World Food Security," *World Watch*, July/August 1998, p. 11.

31. Population of Beijing and Tianjin from China Internet Information Center, op. cit. note 14; Hai River basin from Dennis Engi, China Infrastructure Initiative, Sandia National Laboratory, at www.igaia.sandia.gov/igaia/China/China.html.

32. Author's estimate.

33. Ma Jun, *China's Water Crisis* (Norwalk, CT: EastBridge, 2004), pp. 148–56.

34. Ma, op. cit. note 29.

35. Ibid.

36. Yang and Zehnder, op. cit. note 28, p. 85.

37. World Bank, op. cit. note 28, p. vii.

38. USDA, op. cit. note 1.

39. Figure 8–3 from ibid.

40. Figure 8–4 from ibid.

41. Business Daily Update, "Food Prices Up 7.9% in March," *Financial Times Information Limited*, 20 April 2004.

42. "China Delegation Bought More than 500,000 MT US Wheat—Traders," *Dow Jones Newswires*, 19 February 2004; China National Grain and Oils Information Center, *China Grain Market Weekly Report* (Beijing: 16 April 2004); "Vietnam Says China Seeks 500,000 T Rice, Paddy," *Reuters*, 31 August 2004.

43. "State Raises Rice Prices," op. cit. note 3; Susan Cotts Watkins and Jane Menken, "Famines in Historical Perspective," *Population and Development Review*, December 1985.
44. "State Raises Rice Prices," op. cit. note 3; USDA, op. cit. note 1.
45. Figure 8–5 from USDA, op. cit. note 1; International Monetary Fund, *International Financial Statistics*, electronic database, viewed 2 September 2004.
46. USDA, op. cit. note 1.
47. Ibid.
48. Xie Wei and Christian DeBresson, "China's Progressive Market Reform and Opening," at www.unido.org/userfiles/hartmany/IDR-01_DebCHINA-part-I-FINAL-without-introd.pdf, United Nations Industrial Development Organization, 2001; USDA, op. cit. note 1.
49. Arthur Kroeber, "Wanted: Property Rights for China's Farmers," *Financial Times*, 11 March 2004.
50. Support prices from Fred Gale et al., *China Grain Policy at a Crossroads*, Agricultural Outlook (Washington, DC: USDA, Economic Research Service (ERS), September 2001); Hsin-Hui Hsu and Fred Gale, coordinators, *China: Agriculture in Transition* (Washington, DC: USDA, ERS, November 2001); "Unprecedented State Subsidy," op. cit. note 20.
51. FAO, op. cit. note 21.
52. China Internet Information Center, op. cit. note 14.
53. U.S. Embassy, op. cit. note 8.
54. U.S. Embassy, *China Adopts Law to Control Desertification* (Beijing: November 2001).

Chapter 9. The Brazilian Dilemma

1. U.S. Department of Agriculture (USDA), *Production, Supply, and Distribution,* electronic database, at www.fas.usda.gov/psd, updated 13 August 2004.
2. Ibid.
3. Ibid., updated March 2004.
4. U.N. Food and Agriculture Organization (FAO), *The State of*

Food Insecurity in the World 2002 (Rome 2002); United Nations, *World Population Prospects: The 2002 Revision* (New York: February 2003).

5. FAO, *The State of Food and Agriculture 1995* (Rome: 1995), p. 175.
6. Marty McVey, Phil Baumel, and Bob Wisner, "Brazilian Soybeans—What is the Potential?" *AgDM Newsletter*, October 2000; FAO, *FISHSTAT Plus*, electronic database, viewed 13 September 2004; Peruvian anchovy industry from Lester R. Brown and Erik P. Eckholm, *By Bread Alone* (New York: Overseas Development Council, 1974), pp. 155–57; soybean prices from International Monetary Fund, *International Financial Statistics*, electronic database, viewed 2 September 2004.
7. Philip M. Fearnside, "Soybean Cultivation as a Threat to the Environment in Brazil," *Environmental Conservation*, 7 January 2000, pp. 23–38; USDA, op. cit. note 1.
8. McVey, Baumel, and Wisner, op. cit. note 6; Figure 9–1 from Ricardo B. Machado et al., *Estimativas de Perda da Area do Cerrado Brasilero*, technical paper (Brasilia: Conservation International: unpublished, July 2004).
9. Kenneth Cassman, discussion with author, 20 September 2004.
10. McVey, Baumel, and Wisner, op. cit. note 6; Randall D. Schnepf, Erik N. Dohlman, and Christine Bolling, *Agriculture in Brazil and Argentina* (Washington, DC: USDA, Economic Research Service: 2001).
11. Schnepf, Dohlman, and Bolling, op. cit. note 10.
12. Figure 9–2 compiled from USDA, op. cit. note 1.
13. Figure 9–3 complied from ibid.
14. Figure 9–4 compiled from ibid.
15. Schnepf, Dohlman, and Bolling, op. cit. note 10, p. 37; USDA, op. cit. note 1.
16. USDA, "Brazil: Soybean Expansion Expected to Continue in 2004/2005," at www.fas.usda.gov/pecad/highlights/2004/08/Brazil_soy_files/index.htm, 16 August 2004.
17. Marty McVey, "Brazilian Soybeans—Transportation Problems," *AgDM Newsletter*, November 2000.

18. Ibid.
19. Ibid.
20. USDA, op. cit. note 1; Melissa Alexander, "Focus on Brazil," *World Grain*, January 2004.
21. USDA, op. cit. note 1.
22. Ibid.; Vania R. Pivello, "Types of Vegetation," Embassy of Brazil in the United Kingdom, at www.brazil.org.uk/page.php?cid=283&offset=0, viewed September 2004.
23. USDA, op. cit. note 1; Pivello, op. cit. note 22.
24. USDA, op. cit. note 1.
25. Ibid.
26. McVey, op. cit. note 17; Schnepf, Dohlman, and Bolling, op. cit. note 10; Figure 9–5 compiled from USDA, op. cit. note 1.
27. Livestock exports from USDA, op. cit. note 1, updated 18 March 2004; livestock production data from FAO, *FAOSTAT Statistics Database*, at apps.fao.org, updated 24 May 2004.
28. Figure 9–6 compiled from FAO, op. cit. note 27; beef production described in David Kaimowitz et al., *Hamburger Connection Fuels Amazon Destruction* (Jakarta, Indonesia: Center for International Forestry Research, April 2004).
29. Figure 9–7 compiled from USDA, op. cit. note 1, updated 18 March 2004; price rise in Kaimowitz et al., op. cit. note 28; FAO, op. cit. note 27.
30. FAO, op. cit. note 27.
31. USDA, op. cit. note 1, updated 18 March 2004.
32. Author's calculation.
33. Population from United Nations, op. cit. note 4; Dominic Wilson and Roopa Purushothaman, *Dreaming With BRICs: The Path to 2050* (New York: Goldman, Sachs & Co., 2003).
34. USDA, op. cit. note 1; FAO, op. cit. note 27; Alexander, op. cit. note 20.
35. USDA, op. cit. note 1.
36. Ibid.
37. Ibid.

38. FAO, op. cit. note 5.
39. USDA, op. cit. note 1.
40. Fearnside, op. cit. note 7.
41. McVey, Baumel, and Wisner, op. cit. note 6; USDA, op. cit. note 1.
42. Fearnside, op. cit. note 7, p. 23.
43. USDA, *The Amazon: Brazil's Final Soybean Frontier* (Washington, DC: 2004).
44. William F. Laurance et al., "The Future of the Brazilian Amazon," *Science*, 19 January 2001, pp. 438–39.
45. Kaimowitz et al., op. cit. note 28, p. 5.
46. Rebecca Lindsay, "From Forest to Field: How Fire is Transforming the Amazon," NASA Web site, at earth observatory.nasa.gov, 8 June 2004.
47. Eneas Salati and Peter B. Vose, "Amazon Basin: a System in Equilibrium," *Science*, 13 July 1984, pp. 129–38.
48. Ibid.
49. Ibid.; William F. Laurance et al., "Deforestation in Amazonia," *Science*, 21 May 2004, p. 1109.
50. Kaimowitz et al., op. cit. note 28.
51. Conservation International, "The Brazilian Cerrado," at www.biodiversityhotspots.org, viewed 10 September 2004.
52. Kaimowitz et al., op. cit. note 28, p. 5.
53. Ibid.
54. Laurance et al., op. cit. note 44.

Chapter 10. Redefining Security

1. U.S. Department of Agriculture (USDA), *Production, Supply, and Distribution*, electronic database, at www.fas.usda.gov/psd, updated 13 August 2004.
2. Ibid.; world grain and soy prices from International Monetary Fund (IMF), *International Financial Statistics*, electronic database, various years.

3. United Nations, *World Population Prospects: The 2002 Revision* (New York: 2003).
4. Anita Katial-Zemany and Rosida Nababan, *Indonesia Oilseeds and Products: Palm Oil Update 2003* (Jakarta: USDA, Foreign Agricultural Service (FAS), 1 December 2003); R. Hoh, *Malaysia Oilseeds and Products Annual 2003* (Kuala Lumpur: USDA, FAS, 17 March 2003); for discussion of Brazil's cropland potential, see Chapter 9.
5. For more information on water, see Chapter 6; for temperature and crops, see Chapter 7.
6. U.N. Food and Agriculture Organization (FAO), *FISHSTAT Plus*, electronic database, viewed 13 August 2004.
7. Kenneth Cassman, Professor and Head of Department of Agronomy and Horticulture, University of Nebraska, letter to author, 7 May 2004; Thomas R. Sinclair, "Limits to Crop Yield?" in American Society of Agronomy, Crop Science Society of America, and Soil Science Society of America, *Physiology and Determination of Crop Yield* (Madison, WI: 1994), pp. 509–32.
8. USDA, op. cit. note 1; Lester R. Brown, *Who Will Feed China?* (New York: W.W. Norton & Company, 1995).
9. United Nations, op. cit. note 3.
10. USDA, op. cit. note 1.
11. "Wheat Board Pulls Out of World Market," *Canada Press*, 6 September 2002; "Drought Threat to Australian Summer Crops," *Financial Times*, 27 November 2002; Michael Byrnes, "Australia Drought Further Cuts Crops," *Planet Ark*, 23 October 2002; "The Spector of Starvation," *New York Times*, 15 June 2002; "Poland Implements Grain Export Fees," Dow Jones & Company, Inc., 11 September 2003.
12. "Russian Government Sets Grain Export Duty for Period Until May 1," *ITAR-TASS News Agency*, 15 December 2003.
13. "Vietnam Says China Seeks 500,000 T Rice, Paddy," *Reuters*, 31 August 2004.
14. USDA, op. cit. note 1.
15. Ibid.; prices in IMF, op. cit. note 2.
16. USDA, op. cit. note 1.

17. United Nations, op. cit. note 3; U.S. Census Bureau, Foreign Trade Statistics, "Trade: Imports, Exports and Trade Balance with China," at www.census.gov/foreign-trade/balance/c5700 .html, updated 10 September 2004

18. Agricultural production databases at USDA, op. cit. note 1, and FAO, *FAOSTAT Statistics Database*, at apps.fao.org, updated 24 May 2004.

19. USDA, op. cit. note 1; IMF, op. cit. note 2; Lester R. Brown and Erik P. Eckholm, *By Bread Alone* (New York: Overseas Development Council, 1974), pp. 69–72.

20. FAO, *The World Food Summit Goal and the Millennium Goals*, Rome, 28 May–1 June 2001, at www.fao.org/docrep/ meeting/003/Y0688e.htm; FAO, *The State of Food Insecurity in the World 2002* (Rome: 2002), p. 4; grain stocks from USDA, op. cit. note 1.

21. USDA, op. cit. note 1; United Nations, op. cit. note 3.

22. USDA, op. cit. note 1.

23. Yacov Tsur et al., *Pricing Irrigation Water: Principles and Cases from Developing Countries,* (Washington, DC: Resources for the Future, 2004), p. 219.

24. Intergovernmental Panel on Climate Change, *Climate Change 2001: The Scientific Basis. Contribution of Working Group I to the Third Assessment Report of the Intergovernmental Panel on Climate Change* (New York: Cambridge University Press, 2001), pp. 102–03.

25. Qu Geping cited in "China Adopts Law to Control Desertification," report from U.S. Embassy in Beijing, November 2001, at www.usembassy-china.org.cn/sandt/desertification_law .htm, viewed 23 September 2004; U.N. Environment Programme, cited in GEF-IFAD Partnership, *Tackling Land Degradation and Desertification* (Washington, DC: July 2002).

26. For more information on the consequences of desertification, see Chapter 5.

27. U.S. farm program from USDA, *Agricultural Resources and Environmental Indicators 1996–97* (Washington, DC: July 1997), pp. 255–327.

28. European percent total cropland reserve is author's estimate based on official 10-percent reserve allowance in European

Union's Common Agricultural Policy; Conservation Reserve Program from USDA, Economic Research Service, *Agri-Environmental Policy at the Crossroads: Guideposts on a Changing Landscape*, Agricultural Economic Report No. 794 (Washington, DC: January 2001), p. 16.

29. Grainland expansion from USDA, op. cit. note 1, and from historical data in Worldwatch Institute, *Signposts 2002*, CD-Rom (Washington, DC: 2002).

30. For more information on plant breeding, see Chapter 4.

31. Lester R. Brown, "Paving the Planet: Cars and Crops Competing for Land," *Eco-Economy Update* (Washington, DC: Earth Policy Institute, 14 February 2001); population from United Nations, op. cit. note 3.

32. Lonnie Ingram, "Grand Challenge for Renewable Energy from Biomass," *Florida Center for Renewables*, at fcrc.ifas.ufl.edu/LOI%20Message.htm, viewed 23 September 2004; food-population equivalent is author's calculation based on world grain production from USDA, op. cit. note 1, and on world population from United Nations, op. cit. note 3.

33. Prices from IMF, op. cit. note 2; relationship between ethanol production and prices in Joseph DiPardo, *Outlook for Biomass Ethanol Production and Demand* (Washington, DC: U.S. Department of Energy, Energy Information Administration, July 2000), p. 4.

34. U.N. Population Fund, *State of World Population 2004* (New York: 2004), p. 7.

35. Water use from Peter H. Gleick, *The World's Water 2000–2001* (Washington, DC: Island Press, 2000), p. 52.

36. Potential *cerrado* cultivatable area from Marty McVey, Phil Baumel, and Bob Wisner, "Brazilian Soybeans—What is the Potential?" *AgDM Newsletter*, October 2000.

Index

About the Author

Lester R. Brown is President of the Earth Policy Institute, a nonprofit, interdisciplinary research organization based in Washington, D.C., which he founded in May 2001. The purpose of the Earth Policy Institute is to provide a vision of an environmentally sustainable economy—an eco-economy—along with a roadmap of how to get from here to there and an ongoing assessment of its progress.

The *Washington Post* called Lester Brown "one of the world's most influential thinkers." The *Telegraph of Calcutta* refers to him as "the guru of the environmental movement." In 1986, the Library of Congress requested his personal papers, noting that his writings "have already strongly affected thinking about problems of world population and resources."

Some 30 years ago, he helped pioneer the concept of environmentally sustainable development, a concept he uses in his design of an eco-economy. He is widely known as the Founder and former President of the Worldwatch Institute.

During a career that started with tomato farming, Brown has authored or coauthored some 50 books, which have appeared in some 40 languages. His last book was *Plan B: Rescuing a Planet under Stress and a Civilization in Trouble*.

He is the recipient of many prizes and awards, including more than 20 honorary degrees, a MacArthur Fellowship, the 1987 United Nations Environment Prize, the 1989 World Wide Fund for Nature Gold Medal, and the 1994 Blue Planet Prize for his "exceptional contributions to solving global environmental problems." In 2003, he was awarded the Presidential Medal of Italy and appointed an honorary professor at the University of Shanghai. He lives in Washington, D.C.

Printed and bound by CPI Group (UK) Ltd, Croydon, CR0 4YY

21/10/2024

01777084-0001